PRAISE FOR
AND *T*

'James Woodford...has won one Eureka and two
Michael Daley Prizes. His first book, on the
discovery of the Wollemi Pine (*Wollemia nobilis*),
should put him on the short list for another award.'
Age

'Woodford...has the knack of turning detailed and
complex material into a page-turning botanical
thriller.'
New Zealand Herald

'The partial unravelling of a botanical mystery that
will interest all in the trade, whether professionals
or amateurs, and many general readers. Woodford
gives a good account of the prehistory of the
continent and an easy lesson in botanical
classification.'
Australian

'This is a remarkable story told by a fine writer...a
fascinating scientific thriller skillfully told.'
Gardening Australia

'This is a thriller surpassing all the science fiction
writings...compelling reading.'
Marlborough Express

'All you need to know (along with plenty of
speccy pictures) is in James Woodford's excellent
popular science tell-all, *The Wollemi Pine*. Woodford...
has an obvious passion for the subject, and a
formidable knowledge...What more can I say
except "enjoy".'
Big Issue

THE WOLLEMI PINE

James Woodford was born in New South Wales in 1968, and is a science and environment writer for the *Sydney Morning Herald*. In 1996 he won the Eureka Prize for environmental journalism, and was awarded the prestigious Michael Daley Prize for science journalism in 1996 and 1997. *The Wollemi Pine* has become a national bestseller. James is also the author of *The Secret Life of Wombats*, winner of the Whitley Award for Best Popular Zoology Book, and *The Dog Fence*.

THE

WOLLEMI
PINE

The incredible discovery of a living fossil
from the age of the dinosaurs

JAMES WOODFORD

TEXT PUBLISHING MELBOURNE AUSTRALIA

ILLUSTRATIONS

Grateful acknowledgment is made to the following for permission to reproduce illustrative material used throughout the text and in the picture section: Jan Allen, plate VIII; Jane Francis, plate XI, p. 87; Anna Dawson, p. 19; Michele Frank, p. 25; Keith Herbert, p. 155; Bob Hill, p. 116; Ken Hill, plate IX; Mark Mabin, plate X; David Mackay, p. 46; Mike Macphail, plates V and VI; Palani Mohan, p. 138; Bob Pearce, p. 31; Jaime Plaza, plate XIV, p. 28, p. 168, p. 184, p. 199; Peter Rae, plate IV; Michael Sharp, plates VII and XIII; Rick Stevens, plates I, II, III, p. 8, p. 122, p. 127, p. 131, p. 193; Sue Stubbs p. 211; Haydn Washington, plate XII.

Thanks also to the Royal Botanic Gardens in Sydney for use of the following: plate XIV, p. 46, p. 168, p. 184.

For further information online visit the Wollemi Pine Conservation Club at www.wollemipine.com

The Text Publishing Company
Swann House
22 William Street
Melbourne Victoria 3000
Australia
www.textpublishing.com.au

First published 2000, reprinted 2000 (twice), 2001, 2002, rev. ed. 2002, 2003
This new edition published 2005, reprinted 2006, 2007

Printed and bound by Griffin Press
Designed by Chong Weng-ho
Maps by Norm Robinson
DNA chart on p. 160 based on original by Rod Peakall
Typeset in Centaur by J&M Typesetting

National Library of Australia
Cataloguing-in-Publication data:

Woodford, James, 1968– .

The wollemi pine: the incredible discovery
of a living fossil from the age of the dinosaurs.

New ed.
Bibliography.
Includes index.
ISBN 978 1 920885 48 9.

1. Araucariaceae – Evolution. 2. Conifers – New South Wales. I. Title.

585.309944

This project has been assisted by the Commonwealth Government through the Australia Council, its arts funding and advisory body.

To Rick Stevens,
the eyes for my stories.
His wisdom, patience and unshakable enthusiasm
are an inspiration.

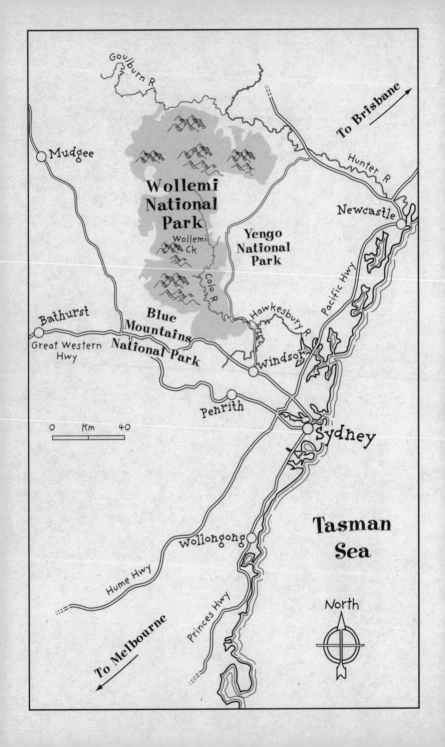

Wollemi pines remained undiscovered for more than 200 years of European occupation of Australia and survived more than 60,000 years of Aboriginal firestick farming. They have a primitive beauty and are among the most conservative and resilient organisms that have ever lived. But this is not the reason for their survival. They survived because of the existence of a remarkable wilderness. Climate change could not fully penetrate its dark and wet interior and, while people have schemed to tame the forests and gorges of Wollemi, none have succeeded.

The location of the pines is a secret and the current intention of the managers of the trees, the New South Wales National Parks and Wildlife Service, is that researchers will be the only visitors ever allowed into the site. Some details in this book have been changed in order to conceal the exact position of the canyon where the trees live. There are good reasons for this. If disease arrived in the colony on the soles of bush-walkers' boots, the botanical find of the century could be wiped out.

There's one more reason to leave the trees in peace—to harm them is punishable with a fine of up to $220,000 or two years' imprisonment.

Contents

Chapter 1

MODERN GONDWANA

On Saturday 10 September 1994 three men arrived on a ledge halfway down a grey rock face in the Wollemi wilderness, about 100 kilometres from the city of Sydney. They rummaged through small bags worn around their waists to get their ropes. Even though it was cold and they were extremely fit the men were gleaming with sweat. The leader of the three was a young National Parks and Wildlife Service field officer named David Noble. His climbing companions were Michael Casteleyn and Tony Zimmerman. The men didn't talk much; they had spent enough time together in similar places to know what each was thinking. The odd smile and the occasional nod of gratitude when something was passed from hand to hand was easier than talking—these three knew that they needed to concentrate.

Noble pulled at the base of a thin coachwood sapling which was struggling to survive in a crack. It was the only suitable anchor point on the ledge. He pulled the small tree several times, using the strength of his whole body, even though he knew from experience that the sapling was adequate to the task. The young tree had sent its roots far into the wall searching for dampness.

Most people would not have thought this spindly tree worthy even of stripping and drying for kindling. Noble slung a rope around its base. Still no-one spoke. None of the three needed to put on his harness because this place was so rugged that one climb soon followed another and while they descended into the gorge it was easier to wear their straps, pitons and carabiners. Now Noble threaded the rope through the clips on his harness, stood on the edge and dropped as if he were a spider on the end of a web. A whizzing sound filled the air for a few seconds until he fell below the level where the forest reached up the cliff. Casteleyn did the same, followed by Zimmerman. The last sign of them was the sight of their rope being retrieved, twisting like a skinny snake in free fall to the men waiting in the rainforest on the canyon floor. Once under the roof of leaves they disappeared from the outside world as if they had never existed.

Wollemi is a green apparition bound abruptly to the north-western edges of Sydney. From my office high in the skyscraper where the newspaper I write for is based I can see across the entire plain upon which the city's four million residents have built their homes. On the western edge of this plain is a bluish rampart stretching north–south as far as the eye can see. This wall is the beginning of the Greater Blue Mountains, the barrier which Europeans took more than a quarter-century to broach after they first settled at Port Jackson in 1788, and which in turn forms part of Australia's Great Dividing Range. This spine of mountains runs down the east coast of the continent and is a fortress-like wall—to its west is merciless weather that creates deserts and to the east a more benign climate replenishing an environmental storage cupboard of the biodiversity of Australian rainforests.

The Blue Mountains contain a million hectares of forest and the Wollemi wilderness, into which Sydney's sun drops at dusk, makes up roughly half of this ring of bush. Its boundary is joined to suburbia just as a shadow is attached to a heel. With its chasms, mazes and rainforests Wollemi remains a mystery to most Sydneysiders, yet it is as large as the city it abuts. Its 'dreadfully rocky and…perilous Gullies' provided prodigious difficulties to early colonial map-makers, and much of it is still a cartographic black hole, mapped mostly from aerial photographs.

Wollemi can also be a malevolent place. People vanish in there never to be seen again; it challenges even experienced walkers and is the kind of environment where aeroplanes crash, bushfires start and escapees hide out. The most notorious of these was the 'Lady Bushranger', Jessie Hickman, who murdered her abusive husband in the 1920s by ramming the leg of a chair through his temple before escaping into the wilderness. From her base in the deep canyons near Nullo Mountain, in the north-west of Wollemi, she stole cattle and made mischief. Jessie was more than 1.8 metres tall, dressed like a man, carried a gun and had a reputation for being able to fight like a 'threshing machine'.

Perhaps Wollemi will remain untamed forever: it now has the extra legal protection of formal wilderness status, which only parliament can revoke. It is now law that most of Wollemi remain free from all commercial pressures and that the only appropriate human access to the area is on foot.

The canyons of Wollemi are part of a massive disturbance to the earth that forms one of Australia's most awe-inspiring landscapes. From the air they resemble a series of bizarre doodles as though someone had carved into the ground in anger. Nearly all of the wilderness's most dramatic features are within what hikers call the 'canyon band'—a six-kilometre-wide area that stretches for nearly 200 kilometres from Katoomba in the Blue Mountains to the Goulburn River National Park. Wollemi is a metropolis of sandstone skyscrapers, every bit as dramatic as Manhattan—in fact if the Empire State Building were dropped into one of the deeper canyons of the wilderness only its mast would be visible. Nobody has ever counted how many pristine water-carved cracks, crevices and gorges it contains but the best estimate of the number of true canyons is around 500. Of these more than 100 are thought never to have been visited by a white person. Some are so impenetrable it may be that no human has ever ventured into them. Veteran hikers know that the Aborigines named this sandstone maze well. Wollemi is believed to derive from the Darkinjung word *wollumnii* which means 'look around you' or 'watch your step'. It is also the name of the longest waterway that runs through the wilderness—Wollemi Creek.

Scientists say that visiting this world of forests, swamps, cliff lines and freezing streams is the nearest thing to seeing the continent when it was a land of dinosaurs rather than of kangaroos and emus. Many of the relict plant and animal species that evolved in the ancient super-continent of

Gondwana, before Australia broke off from Antarctica and began its inexorable movement north, survive in the dungeon-like dankness at the bottom of a Wollemi canyon. It is also one of the last places in the Sydney region where a visitor can see almost a full suite of the marsupial and monotreme fauna which set the continent apart from the rest of the world. Platypuses, echidnas, wombats and koalas all live in the park. (One of its major rivers, the Colo, derives its name from a bastardisation of the Aboriginal word for koala.) It is an original ecosystem which has scarcely changed, not just since Europeans conquered the continent, but for thousands of years before that.

Recognition of this unique environment only started to form in the twentieth century. In the 1930s Myles Dunphy, the founder of the wilderness protection movement in Australia, and his bushwalking friends, began hiking in the region and became increasingly concerned about the likelihood of its being developed. For the next four decades there was a real chance that the Wollemi would be mauled by roadworks, coalmining, forestry or, worst of all, by the construction of a dam to generate hydroelectric power. None of these schemes eventuated, and Wollemi became a national park as recently as 1979. It had taken two centuries for people to come to terms with the reality that the best use of such a place was to do nothing with it whatsoever. This was the strange natural world that David Noble began to explore at the beginning of the 1980s.

Born in 1965, Noble spent his weekdays building walking trails for disabled tourists visiting the Three Sisters—a world-famous triplet of monoliths in the Blue Mountains. On his weekends he deliberately lost himself in the wilderness areas.

Noble is tall and strong and as wiry as a gum tree. He has an old-fashioned face and a reserved manner. He despises the days when he is forced to stay inside and harbours a dislike of those who litter the bush or whose climbing equipment scars the rocks of the Blue Mountains. He is a shy and respectful man who would choose to listen rather than talk and is one of the few people whom the rainforest seems to let pass without a struggle. His body looks like one of those figures that doctors have on their desks, showing the muscles with the skin stripped away. Every sinew is lithe but bulging and he is a renowned bushwalker—the first to reach any destination on a hike. He gives the impression of being the person least likely to crack under physical or emotional pressure and his bosses regard him as a special breed of employee. He is determined to the point of stubbornness. He once said without any irony that his strength was 'visualising how

David Noble, explorer of Wollemi.

something can be done with efficiency and then working steadily and consistently until it is'.

Noble's parents—his father John was a toolmaker, fitter and turner, who also put his hand to working in post offices and general stores, and his mother Olive was a nurse—emigrated to Australia from England when David was two. He describes them as his mentors and inspiration. By the time he was thirteen the family was living in the Blue Mountains. John is now dead but both parents are legendary in the region for their knowledge of plants. The couple spent much of their spare time together at meetings of the local National Parks Association and the Blue Mountains Conservation Society.

Another of Noble's mentors, Wyn Jones, who in the mid-1990s was a NPWS naturalist, remembered David becoming a part of his life in 1992. Both men were on an epic bushwalk from one end of the Blue Mountains to the other. 'That's where my impression of him started,' Jones told me. 'I knew his father quite well and his father had that glint in his eye—faraway but close up—that his son's got.' It was when Jones told me this that I realised that Noble's face and eyes reminded me of an eighteenth-century portrait. He has the same eyes as those men who sailed off the edge of the known world, explorer eyes that are dreamy and sharp and need feeding with new adventures.

When Noble first began walking in northern Wollemi there was little information about the twisted countryside in any guidebooks. He was drawn to the place because it was unexplored, a land without people or rules. Now he mounts thirty expeditions into the wilderness each year and he reckons he has been into 375 of its canyons. Noble always travels in a party of between two and five people. He has done more than 200 times what most people would love to do—choose a name for a place that no other human may have ever seen.

In June 1994 Noble and four friends abseiled into a Wollemi canyon that they had never before visited. As they travelled up the gorge eucalypts gave way to a dense rainforest. The canopy was dominated by towering coachwoods and sassafras. Towards the end of the day's walk they abseiled thirty metres down a rock wall clothed in ferns and orchids, clambered over a three-metre waterfall and arrived at a dry spot under a ledge that the walkers noted as a possible campsite for future visits—a good place to remember in an emergency.

Climbers and walkers in Wollemi call these spots camp-caves. Some are barely more than a dry indentation. They are a welcome place to wedge a sleeping-bag when the wind howls across the park. Each time I have been into Wollemi I have heard these canopy winds roaring through the wilderness. At their worst, storm fronts are visible as waves travelling across the roof of the forest as if the crowns of the trees were one enormous flapping tarpaulin. In such conditions a 400-metre-thick roof of rock can be very comforting. Other camp-caves are like giant cathedrals, with beaches of pristine white sand under their overhang, big enough to sleep the largest parties of campers. Their locations are usually closely guarded secrets.

Three months later, on 10 September, accompanied by

Casteleyn and Zimmerman, Noble left for a second trip into their newly discovered canyon. The two men had accompanied Noble on dozens of walks in the north of Wollemi. Casteleyn was an engineer at Bathurst City Council and Zimmerman a storeman at a supermarket in the Blue Mountains.

The party was carrying sixty metres of rope, an abseiling harness each, topographic maps made from aerial photographs, a compass, a torch, a box of matches, lunch and afternoon tea. The men were wearing old grey tennis shoes, football shorts and t-shirts. Around their waists were bumbags to store their modest list of equipment. They took no tents, no sleeping-bags and no warmer clothing because their aim was to get in and out in a day. Most campers walk to get somewhere pleasant, light a fire and sit around shooting the breeze. For Noble's party the pleasure *was* the walk.

Unlike most climbers Noble uses six-millimetre rope— less experienced canyoners use nine or eleven. The breaking strain of the six-millimetre when new is about 500 kilograms, but after wear and tear it can snap if suddenly tugged with a strain equivalent to 300 kilograms. Shock loading can easily apply that kind of stress if a climber is not careful.

Even the best walkers who have hiked with Noble speak in awed tones of his fitness and pace, his ability to climb and his navigation skills. To travel as Noble does through the Wollemi wilderness is a form of genius. It is not to learn by rote every nook and cranny or to plan an expedition as if the forest could be conquered, but a matter of sensing the patterns in the landscape so that every route, even if it has never before been travelled, can become predictable.

Noble has one fundamental rule in the Wollemi—the rule of thumb. Exploring creeks and canyons where no-one else has been, he is regularly forced to commit himself and his team to routes of no return. Often they must attach their abseiling equipment to tiny trees which non-walkers would hardly call saplings. The rule of thumb is the minimum diameter of a tree safe to use as an anchor—a trunk thicker than this can be trusted as you lean over the lip of a cliff above a drop into a deep canyon. Noble also slots stones into cracks and uses them as anchors. This confidence in his ropes means he is able to get into places others would never go and leave barely a trace of his presence.

The rule of thumb only failed Noble once. It was on a trip with Wyn Jones. Two people in the party had already gone down the cliff and Noble forgot to tell Jones, who is a bigger man than Noble, that he shouldn't lean back too far. Jones dropped heavily, the anchor point gave way and he fell—fortunately only a few metres.

Noble and his friends tackled the new canyon from the opposite direction when they entered it for the second time. They started their walk upstream, before heading up the escarpment and then into a tributary of the creek they were exploring. In this tributary were three abseils—ten metres, then twenty, then six. In spite of it being early spring, with the temperature hovering around ten degrees Celsius, the three walkers had to strip down twice to swim through two-metre-deep pools. The water was freezing, seeping into the main waterway from aquifers deep within the sandstone or gurgling down from other canyons that never experience direct sunshine.

The party climbed a wall on the other side of the creek, went up a canyon on that side, then continued down the original canyon. Noble was walking in front of the others. At that point things seemed strange to him and he stopped—unusual in itself. He had broken through the dense vegetation and found himself in an area that was slightly more open. 'I had seen thousands of these gullies,' Noble recollected, 'and it looked totally different to the coachwood and sassafras rainforests that you normally find.' Ahead of him big, strange trees were growing. Their bark was weirdly bubbled and reminded Noble of Coco Pops breakfast cereal. Below the trees were mounds of debris, as if someone had raked the floor of the forest into piles of compost. The foliage on the ground was brown and distinctive and Noble later remembered thinking that he had not seen anything like it. His eyes moved from the mounds to the trunks and followed their massive cylindrical shapes skyward until his neck was craned back. Odd, he thought.

'Maybe I was the first one to see them,' he remembered. 'I picked up the leaf litter and said, "That looks a bit different." I went over and had a look at the tree it came from.' On the spur of the moment he decided to souvenir some leaf samples. 'Righty-ho, I'll just take a little cutting of it,' he thought. He broke off a piece of juvenile foliage about as big as his hand and stuffed it into his bumbag.

The party then explored a tributary of the main creek before walking back to the peculiar trees. It was mid-afternoon. The three men drank from the creek, filtered by the massive root systems and leaf litter, and Noble had a snack of fruit cake that his Mum had made. The walkers then headed further down the creek. A quarter of an hour later they were at the camp-cave they had found three months before. They climbed up yet another rainforested, almost impenetrable, tributary and after dark arrived back at where they had begun their walk. At the time Noble did not know it but the grove of strange trees he had found consisted of a mere twenty-three adults. The fragment of foliage that he collected was perhaps the first to leave the canyon since the beginning of the last ice age 40,000 years ago. It stayed in his bag, temporarily forgotten, for the next forty-eight hours.

By the time Noble retrieved his cutting it was dried out and battered. He showed it to his father, who had never seen anything like it. The next day he took the sample to his mentor at work, Wyn Jones. What sort of plant did it come from, asked Jones, fern or shrub?

'It's a bloody big tree,' said Noble.

While Jones was recounting this story to me he was delicately rummaging through a plastic tupperware container, searching for the remains of the branch that Noble had brought back to civilisation. His home was a friendly shambles: his partner had just had triplets, one of whom is called Milo after the great Australian wilderness conservationist Milo Dunphy, the son of Myles. The evidence of all Jones's passions were on display—his music, his plants and his photographs of the Blue Mountains.

When Noble brought in his find Jones was fifty-two years old, a naturalist with twenty-five years' experience in identifying plants and an inveterate hater of cities and bureaucracies. His background is in agricultural science and in 1975 he joined the Forestry Commission, which managed the state's production forests. In 1981 he left, frustrated at not being able to push harder for a stronger conservation ethic. He then became the NPWS central region senior naturalist, responsible for the slab of New South Wales stretching from Wollongong in the south to Taree in the north and Parkes in the west.

In the mid-1980s Jones discovered he had chronic fatigue syndrome and was forced to retreat and refocus. He decided to concentrate on the endangered species of the Blue Mountains. He recalled that this period of ill health was at first 'shattering' but ultimately a godsend. Away from the head office of his bureaucracy he was able to concentrate on the ecology of the sandstone escarpments and canyons behind Sydney. In 1992 he travelled to the United States and in a desert in its far west he put himself through an ancient Indian isolation ritual, which he described as a way of becoming at one with the land. Jones is a passionate man.

Jones also visited the Mecca for any human interested in the study of conifers—the bristle cone pines of the western deserts in the United States. These are more than 4000 years old. This experience finally led him to organise the Great Blue Mountains Heritage Walk, the two-and-a-half-month-long trek where he met Noble, from the northern tip of Wollemi to the southern extreme of the Blue Mountains in Mittagong. If you haven't bushwalked with Jones then it is impossible to understand him. He loathes talking on the telephone and refuses to talk straight unless he wants to. Friendship and shared wilderness experiences are indivisible for this non-conformist who fights battles with bureaucrats and can dish out frustrating tests of loyalty to those around him. But David Noble is the kind of individual for whom Jones has time.

'The thing about Dave is that he's one of those observers who are really aware,' Jones commented. 'It was something more than a physical awareness and I knew I could train him. I had actually said to Dave sometime when we were doing a canyon that I reckon there's a bloody big tree out there. It was a half-joke that I had had with a few people for a few years. It was such a big area and I knew how bushwalkers are when they're there. They don't usually visit the nooks and crannies where things might hide and secondly they're not usually trained. The sort of observant people who go there are few and far between.'

'So this thing landed on my desk,' Jones continued. 'I didn't know what it was. I thought I did at first. I think I said it was a fern, but when you say something you immediately question it. Well, I do. It's like working on a musical score—notes crop up and you eliminate some. Certainly it was in the group of so-called lower plants.'

These 'lower plants'—algae, ferns, cycads, conifers—lived on earth before flowering plants evolved and were among the first organisms to colonise dry land. One of Australia's most primitive groups of plants—the cycads—ran through Wyn's mind. 'Not for one minute did I think it was a flowering plant and that was critical. If he'd said it had bloody flowers on it I would have been extremely surprised. He brought in a fairly dried, juvenile leaf specimen with three stages of growth. Then he said it had bubbly bark. Coco Pops bark came soon after.' Jones continued to run through the possibilities. 'This thing was a bloody great tree and it was fern-like. So I thought it was the biggest bloody fern in the world. That's not stupid because some of the old fern-like plants were huge trees as big or bigger than any we've got now.'

Jones searched through books on Australian flora and speculated that the foliage retrieved by Noble might be a type of cycad, ubiquitous in the Australian bush, called a *Macrazamia*. He took the material to the Royal Botanic Gardens Identification Unit. It was then passed on to Ken Hill, a senior botanist at the gardens. Hill could not identify the sample either, though he did suggest it might be from a Chinese conifer called *Cephalotaxus*. Its foliage is almost identical to the juvenile leaves of the tree that David Noble had found. Even so, Jones had a feeling that Hill was wrong.

Jones began to plan a trip to the canyon with Noble, who refused, however, to head straight back to a place he had already visited. It was decided that another canyon would be explored on the way to the weird trees. Jones and Noble had visited the Wollemi wilderness around a dozen times together without generating much interest. But the enticement of a potentially important find soon attracted a party of six for the expedition. 'If it had been a bloody big shrub no-one would have been nearly as interested in going out on that trip,' Jones recalled.

'I reckon there's a bloody big tree out there.' Wyn Jones had a hunch that the wilderness might be hiding something extraordinary.

They arrived in the canyon on 15 October 1994, about an hour before dusk, which in the bowels of Wollemi is getting pretty close to dark. No-one had brought equipment to stay overnight. When Jones entered the grove he saw leaves in the creek and looked up. 'I saw it straight away,' he said. 'My head was in that bloody canopy and there it was. It was quite different from all the other rainforest trees around it: its shape and the type of density of the foliage and the patterning of the branches. I took a photograph straight away of that particular tree.'

'Wyn was running around quite frantically,' Noble said. 'I got the impression that this was different from anything he had seen before. I realised then that he was onto something new.'

Jones used both his video and stills cameras and began stuffing a few extra branches into his backpack. 'I knew I needed more material, sexually active material,' he said. He gathered more of the lower leaves—green ones this time—and pollen cones. 'We got bubbly bark. That was critical. The bubbly bark was probably the single most stunning feature of it.'

Jones also noticed that there had been a fire in the stand of trees. He saw that there had been a tree fall and that, whatever these massive organisms were, they were multi-stemmed. He observed at least two different leaf types, totally flat juvenile leaves and twisted adult leaves. He also noticed that seedlings were scattered through the forest which meant trees were reproducing.

The puzzle had deepened. Jones knew he would have to come back and somehow get to the tops of the trees to investigate what looked like big knobs on the upper end of the branches. He assumed that these were the female cones, which would be critical to identification. 'There were shapes. There was a dark-green silhouette against the sky and all I could see were the shapes. There was something telling me that we had to be very careful with this one because if it was genuinely a rare thing then I needed to sound a warning bell. We were excited.'

The walkers regrouped, left the trees and headed upwards and out of the canyon. Within minutes they were away from the creek and above the rainforest, walking in the thickening gloom back to civilisation.

Chapter 2

ARE YOU MY MOTHER?

Thrilled and confused by seeing the trees in the wild, Wyn Jones sought further help in his quest to identify them. He turned to botanist Jan Allen, a forty-year-old New Zealander whose field work included working on remote islands off the New Zealand coast. In 1994 Allen had been in Australia for two years. A brunette, she often wears her hair back, her features are warm and soft and she is no snob in respect to her science. She met Jones while knocking on doors around Sydney with her CV. Soon after this she found herself soaked through and shivering underneath the giant waterfalls in the highest parts of the Blue Mountains, surveying a remnant patch of *Microstrobos* trees. This endangered, diminutive conifer—its name means 'tiny cones'—has survived the desertification of the continent by hiding in the soaking mists of waterfalls. Apart from a different species in the same genus that grows in the highlands of Tasmania it is found nowhere else in the world. *Microstrobos* is Jones's truest and deepest passion. He knows each tiny patch by heart and can recognise if even a few branches have been broken off in a flood or by vandals. He also judges people by the level of regard they hold for these remarkable little trees. Before he would utter a single word to me about David Noble's find we spent an entire day exploring the *Microstrobos* populations in three different waterfalls.

In the spring of 1994 Allen was finishing off a contract at the University of Western Sydney as a technician in a microbiology lab. She was also on the casual list at the ticket booth of the Mount Tomah Botanic Gardens, in the Blue Mountains. On Tuesday 18 October she came home from work to find Jones talking to her partner Rob Smith, who was curator–manager at the gardens. Jones had with him some of the foliage he had collected on his hike into the gorge and an old male cone, by now mostly disintegrated. Plant experts can easily tell the difference between male cones, which produce the fertilising pollen, and female cones, in which the seeds

Jan Allen: her friendship with Wyn Jones set her on the path of identifying the strange tree David Noble had found.

develop after pollination. Generally pollen cones are long and thin and female cones are round.

That evening these three Blue Mountains plant specialists went down to the Mount Tomah gardens with the cuttings from the wild and walked around its conifer sections. That walk highlights the personal way that Allen approaches her science. She described it to me as being like a story she read to her son in which a little lost duck must ask each animal it meets, 'Are you my mother?' and each animal replies, 'No, I am not your mother.' Finally the lost duckling finds its mother and they all live happily ever after.

Jones and Allen were not so lucky, but from that moment they agreed David Noble's mysterious find was their project. They jealously guarded their progress, from concern about the trees' well-being and to protect their scientific control. From this moment they did not discuss their work with any of the experts at Sydney's Royal Botanic Gardens. If at this point the male pollen cone that Jones now had in his possession had been shown to Ken Hill, the identification of the tree might have been completed in just a few minutes. It would be many months before Hill, who felt embarrassed about his initial guess, would be told that he had not been given all the pieces of the jigsaw.

In order to understand the millions of different types of organisms that inhabit the earth, scientists have divided them into groups. Plants are organised into families, genera and species. A genus—the singular of genera—describes a group of organisms within a family that share common traits. For instance, all gum trees belong to an international family called the Myrtaceae. Myrtaceae is an enormous grouping that in Australia alone includes bottlebrushes, paperbarks, the lilly pilly, turpentine trees and, elsewhere, clove and guava plants. Altogether Myrtaceae is made up of around 100 genera and 3000 species. The genus *Eucalyptus* is the most ubiquitous and famous of the Myrtaceae in Australia, and contains a total of around 900 species. What Jones and Allen were attempting to do as the first step to understanding Noble's discovery was to place this new conifer in a family. Jones had begun to think that it may belong to a group of one of the most primitive of the world's trees—the bizarre and ancient Araucariaceae family.

Wyn and Jan knew that the family Araucariaceae contained two genera—*Araucaria* and *Agathis*. The pair learned from further research that the genus *Araucaria* has nineteen species and the genus *Agathis* twenty. Many members of this family are known as 'monkey puzzles': the Chilean conifer *Araucaria araucana* was named after the country's Araucanos Indians in the 1830s, and monkey puzzle is thought to refer to the fact that the trees are ferociously prickly, making access to them, even for monkeys, difficult. Each tree wears these prickles or spines like a bodysuit that covers trunk, branches and stems. Monkey puzzles are the classic dinosaur-tree family, with their bizarre, sometimes fern-like foliage, their appearance of tropical fecundity, their straight and simple architecture and their imposing dimensions.

A British expert on the Araucariaceae family, Chris Page,

The monkey puzzle trees of Chile—Araucaria araucana—*are so reminiscent of the Cretaceous era that they were used as the backdrop for the documentary* Walking with Dinosaurs.

recently proposed that the monkey puzzles remind us of dinosaurs for a good reason—they evolved to look like them as a means of scaring away herbivores. Page argued at a symposium in 1998 that dead Araucariaceae look like the skeletons of herbivorous dinosaurs—giving any hopeful browsers a serious pang of doubt before they enter such a 'graveyard' for a feast. Page dubbed these dead trees 'palaeo-pseudoscarecrows'. He further observed that some trunks in monkey puzzle forests resemble large reptilian feet, which might give herbivores the impression that carnivores were lurking in the forest. 'Dinosaurs may be extinct, but has anyone told the Araucarians?' he asked.

The family has a reputation for being made up of botanical hermits. They tend to grow in fragmented populations or, like the Norfolk Island pines, happily chug away in isolation. All Araucariaceae are cone-bearing, and all conifers are members of an even bigger grouping—the gymnosperms, plants with 'naked seeds' or, put another way, plants without flowers. Gymnosperms evolved before flowering plants. The sequoias of California, one of which is so big that a hole in its trunk is large enough to drive a car through, do not reproduce with flowers. Radiata pine, an introduced species in Australia cursed by some conservationists, is a gymnosperm. Ricket bush, a native cycad hated by outback pastoralists because cattle who eat it suffer from rickets, is also without flowers. Fewer than 1000 species of gymnosperms exist around the planet—0.5 per cent of the number of flowering plants.

Australia is home to representatives of three of the world's seven conifer families, but the Araucariaceae—including the Norfolk Island, hoop, bunya and kauri pines—is probably the best known. Walk along any street in any city on the east coast of Australia and the chances are that an Araucariaceae will be visible. The south coast towns of New South Wales are infested with the rigidly symmetrical Norfolk

Island pine and there is hardly a beach in Sydney that does not boast one.

The bunya pine, whose seed-cones can reach bomb-like proportions of forty centimetres in diameter and can weigh in excess of eleven kilograms, is the living Australian monkey puzzle with the most ancient lineage. Unlike the rest of the pine, the seed-cone does not possess fierce prickles and is like a pineapple to hold. Its stronghold is the Bunya Mountains of south-east Queensland, near the town of Dalby. During the Jurassic—an epoch which stretches back from 144 million years ago to nearly 210 million years ago—it was far more widespread, extending even into the northern hemisphere. There can be no doubt that this tree would have dispatched countless dinosaurs and other creatures, with a soft thud. Its football-sized cones shower down over the landscape in summer, falling to the ground with a sound identical to the arrival of a bomb that fails to explode. When they land they often hit so hard that the only way to retrieve them is to excavate them with a shovel from their craters. The bunya pine is a brutally tough organism, with awesome defences and an ability to out-compete almost any other tree in its preferred habitat. Bunyas were widely planted in Australian parks yet in many cases proved to be a health hazard and were removed. In 1999 and 2000 the ranger at Cumberland State Forest in suburban Sydney, Prue Bartlett, was forced to close her forest

State Forests ranger Prue Bartlett with a botanical bomb: the seed-cone of a Bunya pine.

for a month at a time because of the danger falling cones posed to walkers.

The world's flowering plants evolved in the last 120 million years but the history of plants stretches back into the earliest days of life on earth. More than 4.2 billion years ago the earth had formed a solid crust but it was not until around 3.9 billion years ago, when the catastrophic bombardment of objects from outer space eased, that life had a chance. Until then the planet was like a galactic punching bag, whose oceans were repeatedly vaporised by cosmic collisions. The earliest possible evidence of photosynthesis anywhere in the world comes from 3460-million-year-old stromatolites from North Pole—its real name!—in Western Australia. Scientists, however, cannot be sure whether this was 'oxygen photosynthesis' as it is known today. It is not until more than 2700 million years ago that there is clear evidence of oxygenic photosynthesis. By the time the ozone shield was established, blocking out most of the sun's ultraviolet radiation, algae ruled the oceans and by 2.2 billion to 1.8 billion years ago these first primitive photosynthesisers had probably made the transition to land.

More than 400 million years ago a lycopod named *Baragwanathia*—related to the primitive family of club mosses—became among the first true plants in Australia. *Baragwanathia* grew as tall as a metre with branches up to five centimetres wide. The first Australian trees were Lycophytes which by more than 360 million years ago had reached heights of over twenty metres. During this period continental Australia is thought to have occupied low to middle southern latitudes but was already embarked on half a billion years of bouncing up and down between the equator and the south polar latitudes.

Around 290 million years ago evidence from Argentina

indicates that conifers, which had evolved from primitive seed ferns, were established in the super-continent Gondwana. But by about 260 million years ago a newly evolved group of gymnosperms—the Glossopterids—dominated the southern hemisphere. So common was this family of trees that it left a legacy in Australia alone of at least 120 billion tonnes of coal. Today Glossopterids are extinct. They were destroyed around 245 million years ago in one of a disconcertingly common series of extinction events that have taken place in earth's history. This wholesale dismissal of life probably occurred because of massive climate change wrought by asteroids and volcanism and evidenced by huge swings in the planet's sea levels.

Araucariaceae appear in the fossil record in the northern hemisphere during the Triassic period—around 245 million years ago. The author of *The Greening of Gondwana*, seventy-four-year-old Mary White, the doyenne of southern hemisphere and Australian popular palaeobotany, says that by the Triassic 'it would have been very much the world as I know it'. She told me this in winter beside the fire in the lounge room of her home on the shores of Sydney Harbour. 'You would have had your marshes and your swamps, your heathlands, your forests. You would have had vegetation that looks like swamp vegetation in swamps today. The fact that the individual plants are no longer there doesn't matter at all. Your first view of the land as you came in on the spaceship and landed 200 million years ago would have looked just like it looks now. The sky and the water would look the same and there would be many things that a modern person would recognise.'

As the Jurassic era dawned, more than 200 million years ago, the Araucariaceae family began to appear not just north of the equator but in the Australian fossil record. This was a time when the continent was sitting at latitudes between 35 and 65 degrees south. The Antarctic coastline today sits at a latitude of 60 degrees south. Australia is now a nation of gum trees and wattles but back then eucalypts were not even a distant prospect on evolution's agenda. Instead the landscape of the continent was cloaked in unending forests of monkey puzzles and other primitive conifers; there were no flowers and no grasses.

By the end of the Cretaceous, the epoch that came to a dramatic close 65 million years ago with the extinction of the dinosaurs, the Araucariaceae in Australia seemed to have hit their straps. An act of cosmic violence, however, was again to strike the planet. The fossil record suggests that the same impact that wiped out the dinosaurs also vaporised the Araucariaceae forests in the northern hemisphere. Since then the family has remained almost exclusively a southern hemisphere resident, with just a few minor northern populations in South-East Asia. Scientists hypothesise that the

Million years ago

1.8	Pleistocene	Most recent ice ages
5.3	Pliocene	First humans
23.8	Miocene	Decline of widespread rainforests
33.7	Oligocene	Diverse vertebrate communities
54.8	Eocene	Final separation of Australia from Antarctica
65	Paleocene	Flowering plants assume a modern form
144	Cretaceous	First Wollemi pines First flowering plants Extinction of dinosaurs
206	Jurassic	First birds
248	Triassic	First dinosaurs
290	Permian	First conifers Early reptiles
354	Carboniferous	Early amphibians
417	Devonian	First insects
443	Silurian	First fish
490	Ordovician	Marine invertebrates

bolide, or massive meteorite, which destroyed the dinosaurs had little impact on the southern conifers because it struck in June—the northern summer. The southern hemisphere Araucariaceae were not only further away from the impact zone but also, because it was their winter, were probably dormant anyway.

If what Dave Noble had found in the canyon was in the family Araucariaceae, Wyn Jones and Jan Allen had no idea where the tree fitted into the family. Allen refused to accept Ken Hill's suggestion that the tree was the Chinese pine *Cephalotaxus*. She began driving around Sydney looking for representatives of the vast array of Araucariaceae species that have been planted in suburban gardens and parks seeking clues in fallen cones. Allen made a list of material she needed from Jones' mystery tree. 'I need to know exactly what the bark looks like,' she told him. 'I need you to scrabble around on the ground. I need you to bring me a female cone. There must be some there because the kauri I had been to visit, the bunya I had been to visit, the hoop pine I had been to visit, all had remnants of the individual segments of the female cones on the ground.'

For the researchers the gap was closing between two highly competitive emotions—disbelief and excitement. Both believed that something amazing had been discovered but they

The Wollemi Pine

also knew that the process of identification might be tortuous. And until they could get one of the female cones that lie tantalisingly visible high in the crown of the trees, their task was impossible.

Allen and Jones put up a weed hypothesis to explain the presence of such a strange tree in the wilderness, and then sought to knock this hypothesis down. 'What if the pines in the canyon were one of the lesser known Araucariaceae from somewhere like New Caledonia, which has the world's greatest diversity of the family? What if seeds had been taken to Queensland by sugarcane workers and one of them had moved further south to grow one of these trees and a yellow-tailed black cockatoo had taken a liking to a cone, had flown over Wollemi National Park and had dropped it?'

'If it was a weed, rather than celebrating it as something new, then the parks service would be under an obligation to remove it,' Allen said.

In an attempt to resolve this speculation, Jones flew in the NPWS helicopter around some of the properties which neighbour the wilderness, on the off chance that he might see some strange Araucariaceae trees similar to those in the canyon. National Parks and Wildlife staff based in the Blue Mountains spend considerable time in choppers as it is often the only way to get around. As a senior naturalist it was routine for Jones to be given access in the use of the aircraft, but his search was fruitless.

On Friday 21 October 1994 there was no NPWS helicopter available. Jones chartered a private aircraft piloted by Jerry Coleman, and David Noble accompanied them. Jones was on a different mission this time. They flew into the canyon where the strange trees lived with Jones hoping to grab a female cone from the passenger-side window of the chopper. Hovering at the bottom of the canyon, with the downdraft sending the foliage into paroxysms of shaking, was perilous not only for the pilot and passengers but also for the stand of trees. After several failed attempts Jones stuck his rubber-gloved arm out the window, grabbed a stem with his hand and snapped off a tiny cone. In later months Jones would use secateurs on the end of a pole to pick off the female cones. This was done more sensibly from the open door of the chopper.

Jones and Allen now had various samples of foliage, a male cone and the precious female cone. 'We looked at all the pieces of the tree that we had,' Allen recalled. 'We said, "Araucariaceae." The male cones are Araucariaceae, the leaves are Araucariaceae. Some evidence points to *Araucaria* and some to *Agathis*. We've got all the information but it's still not helping us and I said to Wyn we won't get any further unless we chop up the cone.'

But the pair, still working without the assistance of any of the leading conifer experts in the country and aware by now that what they had found was in every likelihood an internationally significant botanical treasure, were reluctant to take the next step and split open their hard-won female cone. 'I said it might be able to give us some answers,' Allen recalled, 'because one of the big differences between the two genera in Araucariaceae is that in *Araucaria* the ovule or seed is fused to the seed scales—the individual segments of the seed-cone. In *Agathis* the ovule sits up free on about a forty-five-degree angle attached to just one end of the seed scale. In *Araucaria* the seed is fused into the segment. I said, "If we dissect the cone we might get an answer and we might not," because it was very tiny and I didn't know whether it had been pollinated yet. We thought about it for a few days and then Wyn came around one night and said, "Let's do it." '

It was Tuesday 8 November 1994.

'I got out a breadboard and a carving knife, very unscientific I know, and we cut the cone in half very carefully and we teased out some of the scales and sure enough there was this little white seed sitting up proud, and it was at that point that I said, "Wyn, you can start getting excited now," because that is what seeds in the genus *Agathis* do and most of the other information we had on the structure of the tree points towards the genus *Araucaria*. It's not going to fit into either genus. We made a little chart with three columns, what *Agathis* has, what *Araucaria* has and what our unknown thing has.

When we double-checked everything I said, "Wyn, it doesn't fit, and given the description for *Araucaria* and *Agathis* we can't make it fit. We can erect a new genus and a new species or we can go back and rework the whole of the family Araucariaceae. It clearly isn't going to fit into either genus.'"

A new genus had been added to the plant kingdom.

I wanted to know how this moment must have felt for Jan Allen and over the summer of 1998 I collected an *Agathis* cone from a Pacific kauri in Sydney and kept it on my desk for a few days—just as she and Jones had done. Every few hours I picked it up, marvelling at its hardness and tortoise-shell-like impenetrability. Eventually I took my cone into the kitchen and cut it in half with my penknife. I was expecting the blade to battle cutting through the seed. But as soon as I pushed past the first few millimetres, it sliced like a hot knife cutting butter. Before it was even halfway there was a pungent waft of sap and then the seed-cone, which had given the illusion of being a hard green lump, fell apart. And there, inside the cone of a relative of the tree David Noble had found, just as Allen had seen them, were the white ovules sitting proud and free from the rest of the seed scale.

Chapter 3

WOLLEMIA NOBILIS

Ken Hill is a tall, softly spoken botanist with a Grizzly Adams beard, flecked with a surprising array of autumnal reds and browns. He mostly wears t-shirts and jeans and has the patient manner of a man with a pair of teenage children and who has spent years of researching the world's slowest growing organisms—trees. On 22 November 1994 Allen and Jones took the tree material and a draft paper to Hill at Sydney's Royal Botanic Gardens, confident in the knowledge that what they had been working on was a new genus. With all of the cuttings and cones that had been collected now in front of him, Hill knew that what was sitting on his desk was the most remarkable and botanically significant specimen that he had ever encountered.

To understand this part of the story required a visit to the herbarium and the grounds of the Royal Botanic Gardens. Hill had offered to introduce me to all the members of the family Araucariaceae that were growing there and to tell me everything he knew about their bizarre sex life. The herbarium is situated on one of the world's best pieces of real estate—a headland that begins under the shadow of the political and legal heart of Sydney and ends jutting into the harbour at the Opera House. It was a steamy, rainy January day when we met in his office.

'When I opened up the cone,' Hill recalled, 'I just thought, "This is amazing—this is a new genus. This is big news." I had a chance to pull the cone apart. From the outside the seed-cone looked much more like an *Araucaria* cone than an *Agathis*. Inside it had more of the structure of *Agathis* but the seeds were different. I phoned Wyn the next morning to tell him, "Yes, it is new."'

To show me what the password is to become a member of the Araucariaceae, Hill and I walked in the rain to the conifer section. Some of the dripping wet trees were barely recognisable to the untrained eye as conifers, with their soft rounded leaves and smooth trunks.

Araucariaceae range in height from ten metres to more than sixty. Their leaves are all spirally arranged and their seed-cones are hard and woody. Some have leaves shaped like savage daggers, with points like needles, others such as the Norfolk Island pine have tiny, dense, leathery leaves that form long cylindrical stems. One of my earliest memories as a child is pulling the stem out from the foot-long Norfolk Island pine needles at Manly Beach—like removing a sword from its sheath.

After a few minutes Hill found an elongated Araucariaceae pollen cone that had been soaked by the rain and broke it in half. He pointed at one of the cone's hundreds of pollen sacs, each just a fraction larger than a pinhead. They looked like hundreds of minute squids impaled on tiny sticks. Every monkey puzzle tree, he told me, has the same structure in the architecture of the pollen sac. No matter which genus or species a monkey puzzle belongs to, its pollen cone identifies its place in the botanical kingdom as surely as if someone had carved the word 'Araucariaceae' to its trunk. 'It is the fundamental difference between an Araucariaceae and the other conifer families,' Hill said.

He explained on what basis botanists place some Araucariaceae into the genus *Agathis* and others into the genus *Araucaria*. One of the keys is the female seed-cone. It is in fact the seed-cone which has given *Agathis* its name. *Agathis* is Greek for a ball of thread, and the seed-cones are made up of hundreds of scales packed into a dense resinous ball.

When the cone is broken up a trained botanist can immediately divide the Araucariaceae family using the following rules. If the ovule sits on top of the scale, if one side of the seed has a wing and if the exterior of the cone is smooth then it is an *Agathis*. Its winged seeds spin as they fall—dropping like a helicopter approaching its landing pad. If the ovule is embedded into the scale, if the seeds are not winged and if the leaves are sharply pointed then the tree is an *Araucaria*.

In the cone which Jones took to Ken Hill the seed was free from the scales, winged all around and the cone was spiny like a little round pineapple. As Hill described, the seed-cones of this startling new genus had the external appearance of an *Araucaria*, such as the bunya pine, and the internal structure especially in its sexual organs of an *Agathis* such as New Zealand's kauri pine. Jones and Allen were calling it a Wollemi pine but now the genus needed a scientific name.

Soon after its formal identification Wyn Jones suggested the tree's formal names should commemorate the Wollemi wilderness. Allen told him that it would have to be *Wollemia* because in botany the genus is always a noun of the feminine gender.

Both also wanted to honour Dave Noble so they decided to add the Latin noun form *noblei*. Naming the tree, however, turned into the beginning of a bitter dispute between Allen and Jones on the one hand and Ken Hill on the other. Hill, who did play a crucial role in giving scientific rigour to the taxonomic description of the new genus, was frustrated that he had been excluded from the early identification process. Allen and Jones felt that they had done all the hard detective work and therefore decisions about naming the species should be made by them. Hill was backed by the full weight of two government bureaucracies, Sydney's Royal Botanic Gardens and the National Parks and Wildlife Service. Allen and Jones, although both associated with government agencies, were discovering that the days of the tree being theirs alone were numbered.

Hill, together with his colleagues and superiors at the Royal Botanic Gardens, insisted on changing the species name to *nobilis*, the Latin adjective meaning 'noble'. *Noblei*, they feared, would be bastardised to 'nobbly'. A tense vote was held between Allen, Jones and Hill. Allen voted for *noblei*, determined to see Dave Noble honoured. Jones voted with Hill as part of a deal ensuring that Jan Allen would be allowed to be a joint author of the scientific paper announcing the discovery of the tree to the academic world. Thus the new genus became *Wollemia nobilis* by two votes to one.

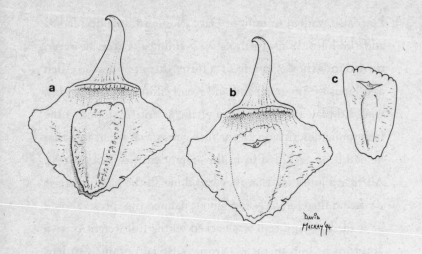

Cut open a Wollemia seed-cone and this is what you will see. On the left (a) the curly tail at the top of the seed-scale complex sticks out of the cone and gives it a prickly appearance. The seed itself is the fingernail-shaped object sitting on the scale. The drawing in the middle (b) shows the scale with the seed detached. And to the right (c) is the seed itself which has a papery wing all around it.

I first began to hear rumours that a weird species of fern had been found in September 1994, just days after Noble brought back the first fragment of foliage from the canyon. Such occurrences are fairly common in Australia and I didn't think too much more of it, though I asked contacts in the National Parks and Wildlife Service to keep me informed. I began to hear gossip through November that something big had been discovered but was unable to find out what. On Sunday 11

December its full magnitude was revealed to me by NPWS staff, who invited me to travel to the Blue Mountains the next day to meet Jones, Noble and Allen. Everybody was very nervous about how the new genus would be publicised. Jones especially was concerned about the ramifications of the story being published at this point. But by this time—and I did not yet know of all the machinations that were taking place—the management of the tree was being wrested out of his control. At first Jones did not even want to show me material from the Wollemi pines. He later returned home and collected a bag of foliage hidden there. It was so fresh that the plastic bag was full of condensation. We agreed that I would write a story which the *Sydney Morning Herald* would not publish until Wednesday 14 December.

On the eve of publication, I went and met Ken Hill for the first time. He had not yet seen the Wollemi pines in the wild. That afternoon I asked him whether it was possible for me to see any of the fossils to which the living Wollemi pines were related. He had none in his office but together we walked across town, through Hyde Park, the green heart of Sydney, across William Street and into the bowels of the Australian Museum. There a 150-million-year-old fossil from an ancient monkey puzzle called *Agathis jurassica* was produced, which to an untrained eye was identical to the green Wollemi pine frond that Hill had brought with him from his office. This particular fossil specimen had, as it happened, been named and described by Mary White. The *Herald* photographer Peter Rae asked me to hold the flash as Hill posed with the fossil and the frond.

I returned to my office and, for the first time, fully briefed my editors about this incredible story. My immediate boss at the time was Bob Beale, one of the finest and most respected science correspondents in the country. Beale spoke to his superiors and in turn impressed upon them how significant this discovery was. I then sat down to write, and I remember clearly the sensation that I would not be able to do the story justice. After I had prepared a draft, Beale and I sat together and produced our finished copy. It was nearly dark when I left the office. I can still feel the sense of excitement that comes along only a few times in a journalist's career when you know, as you head home to bed, that the presses are churning out a story which everyone in your city will either read or hear about when they wake up.

By Wednesday morning that photograph of Ken Hill had been digitised and wired around the planet. Stories about the Wollemi pine ran everywhere from the *New York Times* to the *International Herald Tribune*, from Japan's *Yomiuri Shimbun* to the *Daily Telegraph* in London. A few months later when I was on exchange with an Indian newspaper the two things that my Indian colleagues most wanted to talk about were Australian cricket star Shane Warne and the 'pinosaur'.

Within days all the researchers whose names had been mentioned in the various stories that were published around the world found themselves being swamped with requests from people who wanted either to see or obtain the tree. A letter from an American claimed that he had already begun to construct a special conservatory in which to grow a Wollemi pine: 'A dawn redwood (about twenty-five feet) is thriving in my backyard. And I lust for the Wollemi pine, both to see it and to grow it.' From a father in Japan, apologetic about his English: 'I found a report of a newspaper which a new kind of pine trees had discovered in Australia, and the pine trees are thought that had extincted in the time when the dinosaurs had still lived. And at your botanical park, the buds come out.'

Unsubstantiated rumours immediately began circulating that one conifer collector had offered $500,000 for a Wollemi pine. These rumours would persist over the next few years. At the end of 1999 Ken Hill was at a conifer conference in the UK when he was told that a group of avid German collectors had discovered the location of the pines and was preparing a raid. Extra checks on the grove were ordered by the management team.

On the weekend after the story had broken I dreamed that a scorching forest fire had broken out in the canyon and that I was standing among smouldering charcoal remains—all that was left of the Wollemi pines. It was not such a ridiculous dream. Wollemi is a place where sometimes the best technique for fighting fires is to let them burn for weeks until they run out of fuel.

Only a couple of months after the pines had been discovered a massive blaze broke out in the park. A Wollemi wilderness fire is almost impossible to tackle on equal terms—usually the best firefighters can hope for is to contain the conflagration within the boundaries of the park.

At the time, the National Parks and Wildlife Service was employing a new technique to fight fires called bird-dogging. The name was borrowed from the nickname given to troops in choppers in Vietnam, whose job was to guide bomber pilots to their targets. David Crust, who in the next five years would spend more time with the pines than probably any other person, was the first ranger in New South Wales to be put through bird-dogger training. Instead of guiding napalm runs his task was to direct converted crop-dusters—called air tractors—to the exact location for a water drop. As a bird-dogger, Crust does not fly the helicopter but sits beside the pilot and controls all communications.

I went along to watch him at work. Sitting in the backseat of a chartered helicopter above the gorges of Wollemi—three choppers in total had been called in to fight the fire—we could see a wall of flame moving through the forest in a front a kilometre long. On the ground were teams of NPWS staff, ant-like from the air and armed with rake-hoes. These firefighters had been dropped in by choppers. The *Sydney Morning Herald* photographer with me, Bob Pearce, was hoping to capture one of the air tractors on film as it dumped several tonnes of water on the inferno below. Over the radio Crusty asked the pilot to do a practice run so the photographer would have an idea of where the plane would fly.

The air-tractor pilot's voice crackled over the radio agreeing to the request. Soon we could see his aircraft swing towards us, at first a mere dot against the smoke, then a beast bearing down towards our helicopter, which was still hovering above the fire-front. At the last second the air-tractor pilot remembered that as this was a practice run he had not offloaded his cargo of water and would be struggling to gain the height necessary to get above the gorge wall behind the fire. His plane and our helicopter were on a collision course because of the miscalculation. Our pilot had no choice but to put the chopper into free fall. It plummeted about a hundred metres, leaving our insides in the clouds and the photographer's gear almost plastered to the roof. By a minor miracle the plane missed and the chopper pilot brought his strained aircraft under control. Crusty turned around from the front seat and said, with a strange smile, 'Sorry 'bout that, guys.'

The final, formal act in the discovery of *Wollemia nobilis* was announcing its presence to the scientific world. 'A stand of large trees has recently been discovered in a remote part of the Wollemi wilderness,' begins the paper that was submitted on the day before Christmas and accepted in March 1995 for publication in the scientific journal *Telopea* under the authorship of Jones, Hill and Allen.

'Distinguishing characters of *Wollemia* not shared with either *Araucaria* or *Agathis*,' they declared, 'are the spongy, nodular bark that sheds its outer surface in thin papery scales, and the terminal placement of male and female strobili on first-order branches...The generic epithet recalls the Wollemi National Park, where this plant was found...The specific epithet honours David Noble of the New South Wales National Parks and Wildlife Service, a contemporary explorer of the remote parts of the Blue Mountains, who first discovered this plant.'

Genetically and in evolutionary terms the Wollemi pine appears to sit somewhere between the two genera known to exist prior to its discovery. 'I would have expected *Wollemia* to have been fairly widespread,' Ken Hill told me during a visit to the gardens. 'The Araucariaceae were fairly widespread and I wouldn't be entirely surprised if Wollemi pines were once in the northern hemisphere.' The family has travelled far across two of the greatest barriers that the earth imposes on life—

time and distance—and yet a Wollemi pine and some other members of the family still look like the ferns from which they evolved. 'The Wollemi pine is 300 million years from a fern,' Hill said. 'It's a very different thing from a fern. *Wollemia* is a conifer, it is a seed plant and all the evidence we have is that the seed plants, including the conifers and flowering plants, came from a single ancestor. Wollemi is closer to the dandelions than it is to the fern. They have both come from a common ancestor more recently than ferns have been related to them.'

The find meant that, while New Caledonia has the world's most diverse collection of monkey puzzles, with five species of *Agathis* and thirteen species of *Araucaria*, Australia is the only place with all three genera—*Agathis*, *Araucaria* and *Wollemia*.

There was now a series of urgent questions that needed to be answered about the Wollemi pine. The first question of all was simple: how had such a big tree stayed hidden?

Had anyone come close to finding them before David Noble?

Wyn Jones has gone to extensive lengths to determine whether the Wollemi pine is recorded in local Aboriginal stories and has found absolutely no hint of this. The seeds of monkey puzzle trees like the bunya pine and the hoop pine have been documented as food sources for Aborigines living in Queensland. Trees notched to help Aboriginal climbers can still be seen in the Bunya Mountains National Park. Each bunya pine was the responsibility and property of a particular man—a right passed from father to son. Considering the detailed understanding Aboriginal people have of their country, it is hard to believe that if the Wollemi pine had been known by a tribe that its presence would not have become important. It is possible that there were so few trees and which proved so difficult to access that the Aborigines had no use for them.

There were at least two previous occasions when researchers almost stumbled on the trees. In autumn 1984 Jones was helping Alex Floyd, one of Australia's great rainforest botanists, with a survey in Wollemi. Jones and Floyd did ten days of field work in the wilderness. They flew down the canyons in helicopters and fixed-wing aircraft and went along a tributary of the creek where the pines were eventually found. They got within 200 metres downstream of *Wollemia*.

Even earlier, in 1982, senior ecologist at the Royal Botanic Gardens, John Benson, was hiking in the wilderness and missed finding the trees by a few hundred metres. These two parties of respected botanists, within two years of each other, had come to within half a kilometre of the pine, proving just how remarkable it was that a scientifically unqualified bushwalker ended up making the botanical find of the century.

Benson was, nonetheless, one of the first humans to see the Wollemi pines—a week after the public announcement he was taken to the grove. He, like Jones, first saw them silhouetted against the sky. 'It was like an alien had landed on earth because it was totally out of place with the vegetation of that area,' Benson told me after his visit. 'Of course it looked different from the eucalypt forest, but it also looked different from the rainforest because of the structure of the trees, the shape of the leaves, the bark—everything.' He said the pines gave the impression of sheltering in that canyon as if they had travelled from far away to escape. 'I have to think of the ents in *The Lord of the Rings*—the trees that fought alongside the hobbits and when they got really motivated they had the capacity to pull up their roots and move. These trees,' he concluded, 'retreated to an ever smaller niche as Australia dried out.'

Benson was also strangely reminded of the plastic Christmas trees that many families keep in dusty boxes in the top cupboard. Each year the box is pulled out, a tall straight trunk is slotted into a base and then in just a few seconds the branches are clicked into place. It was the simplicity of the pine's architecture that was so striking.

One of the key officials involved in easing the strained relations between Jones, Allen and Hill was Carrick Chambers, director of the Royal Botanic Gardens. In the first few months of 1995 a memorandum of understanding was signed between the National Parks and Wildlife Service and the Royal Botanic Gardens that divided up responsibilities for the pine and addressed the sensitive subject of how any profits from propagation would be split.

Chambers, an eminent botanist, also became preoccupied in retrieving the results of a study undertaken in the mid-1980s, when he had found Araucariaceae vegetation new to science among 120-million-year-old fossils from Koonwarra in eastern Victoria. The Koonwarra site is truly remarkable—it is thought to have been a small basin beside a larger lake all those millions of years ago—because an entire ecosystem was preserved there, in effect snap-frozen, including uncountable numbers of fish suspected to have died en masse as they were starved of oxygen. Even more startling is the fact that this ecosystem was thriving near the South Pole—the Australian continent had not yet moved north to its present location. Some scientists believe that at that time Madagascar and India were still joined to Antarctica as was Australia. It would have been possible—if perilous given the wildlife—to walk from the South Pole to well beyond the equator. Winter darkness clamped down on this swampy and hostile landscape for months on end: the aurora australis, or southern lights, and the stars may have been the only illumination. In the forests the vegetation was spaced apart to take advantage of these low-light conditions.

Normally where plant fossils are discovered there are no animal remains and vice versa. This is because acid levels in the sediment that preserve plant material are higher than those that preserve animals. But at Koonwarra flea fossils are bountiful, indicating the presence of mammals, and cicadas that look the same as those that disturb the peace today were fossilised. Cockroaches rustled through this landscape, too, and flies, mosquitoes and spiders appear in the stone as though swatted only a few moments earlier. Such richness suggests that in these stultifyingly humid but cool conditions invertebrates would have made life a misery. A small collection of very precious and rare feathers were also found although no-one knows whether these were from a true bird or from one of their ancestors—a dinosaur. From other locations nearby, of a similar age, scientists have learned that this environment hosted swarms of dinosaurs of every variety: carnosaurs, kangaroo-sized *Leaellynasaura* with specially adapted eyes to cope with winter darkness, pterosaurs flying above the canopy and lurching relatives of stegosaurs. This world was a violent one, and a small fossilised *Leaellynasaura* was found, suffering from a chronic bone infection in one of its tibia. It may have received the infection after being attacked; the fossil evidence indicates that the animal lived for up to five years after its immune system began to fight the disease. Other placental mammals—creatures the size of marsupial mice—lived in this unforgiving environment. Among the dinosaur remains found in this area were lower jaws with dentition that indicates the animal may be an ancestor of the northern hemisphere's hedgehog.

Many types of plants were preserved with the notable absence of flowering plants, which means that the deposits were laid down either before flowers had evolved or before they had reached that part of the prehistoric world. Koonwarra 120 million years ago was a forest dominated by ginkgoes—a tree represented today by a single species first seen by western botanists in China in the eighteenth century. Leaves from Koonwarra's ginkgoes fell in huge autumnal beds and these too have been preserved.

Chambers and his colleague Andrew Drinnan published the results of their discoveries at Koonwarra including drawings and photographs of the material. They did not, however, ascribe a scientific name to the fossilised plant they had found but they clearly stated it belonged to the family Araucariaceae. Almost as soon as he learned of the discovery of the Wollemi pine, Chambers became intrigued by the possibility that it and his mystery plant were one and the same thing. The foliage that Jones had brought into the gardens looked like the fossilised vegetation Chambers and Drinnan had collected. In early 1995 Drinnan was working in Melbourne so Chambers flew down from Sydney with some Wollemi pine material. At Drinnan's laboratory Chambers asked him to bring out the extra fossils they found in Koonwarra. The match was so exact that Drinnan exclaimed, 'Jesus, Carrick, our drawings were upside down.' What they had done was draw the ovule attached to the wrong end of the segment.

'If Andrew and I, when we discovered the fossils,' Chambers told me, 'had said that one day the fossil may well be found alive we would have been drummed out of science.'

There could no longer be any doubt of the truth of Chambers' remark on the significance of the discovery. 'This is,' he said, 'the equivalent of finding a small dinosaur alive on earth.'

Chapter 4

A LIVING FOSSIL

We are all living fossils, a term coined by Charles Darwin to describe 'remnants of a once preponderant order'. Every living thing has an unbroken ancestry stretching back billions of years and very many living species, including humans, have a fossil record millions of years old. But the label 'living fossil' has a special meaning to science and to the public, a phrase for those incredibly rare organisms with a good record in the rocks but which at some point in the distant past have disappeared from the scientific radar only to reappear in some little-explored part of the planet. Often the romance surrounding these stories is heightened by the hiding places of living fossils—deep ocean trenches, dark and wet rainforests or high inaccessible mountains. They are nature's lost citizens. In his book *Life: An Unauthorised Biography*, Richard Fortey perfectly sums up living fossils. 'Of course living animals are not identical to their fossil forebears, and what is meant is that these are survivors from earlier worlds. Often it is implied that there used to be many more of their kind, and now they alone linger on, the last of their persuasion. There is a covert implication that these survivors are like members of the Flat Earth Society—they once had their day, but are now to be cherished chiefly for their eccentricities. Did these animals survive by luck or virtue?'

Two astonishing twentieth-century discoveries of living fossils preceded the Wollemi pine. The first was the coelacanth, a primitive fish which had barely changed in 385 million years. On 23 December 1938 Marjorie Courtenay-Latimer, the thirty-two-year-old museum curator in the town of East London, north-east of Cape Town, South Africa was summoned by a dock-worker when the trawler *Lerine* pulled into port. On its deck she noticed a blue fin protruding beneath a pile of rays and sharks. She pushed those aside and, as she wrote later, saw, 'the most beautiful fish I had ever seen. It was five foot long, a pale mauvy blue with faint flecks of whitish spots; it had an iridescent silver-blue-green sheen all over'. Inspecting it more closely she saw the fish 'was covered in hard scales, and it had four limb-like fins and a strange little puppy-dog tail'.

She persuaded a taxi driver to allow her to transport the stinking carcass back to the museum. At first the strange fish was dismissed as a common rock cod but Courtenay-Latimer decided to send a sketch of it to Dr J. L. B. Smith, an amateur ichthyologist, who acted as the unofficial curator of fishes for the smaller museums on Africa's south coast. He realised that the creature trawled up from the deep had until then been known only from fossils. He sent Courtenay-Latimer a now famous cable on 3 January 1939: 'MOST IMPORTANT PRESERVE SKELETON AND GILLS FISH DESCRIBED.'

The coelacanth had been discovered and was an international story within hours of a local newspaper being allowed to photograph its remains. At first the bizarre creature was thought to be the direct descendant of the tetrapods—the land-living animals with four limbs, including humans. The fish has paired fins which move in a similar fashion to our arms and legs and much was made of the suggestion that humans might be able to trace their ancestry back to the coelacanth. It is now thought that another ancient creature, the lungfish, is its closest living relative and it is not at all conclusive that tetrapods are descended from Courtenay-Latimer's find.

On the day that the Wollemi pine's discovery was announced, almost every paper in the world compared it to another living fossil—the dawn redwood. This was probably the plant kingdom's first international overnight sensation. It is a giant tree, related to the sequoias of California, which in stature are one of the most dramatic expressions of life that have ever existed. Its scientific name is a tongue twister—*Metasequoia glyptostroboides*—and like *Wollemia* the dawn redwood is a conifer.

On 25 March 1948 the *San Francisco Chronicle* ran an eight-column headline: '100,000,000-YEAR-OLD RACE OF REDWOODS—SCIENCE MAKES A SPECTACULAR DISCOVERY.' The byline under the story belonged to Dr Milton Silverman, the science editor at the *Chronicle* and a member of the first western expedition to confirm the existence of the dawn redwoods, which were common all around the northern hemisphere when dinosaurs ruled the earth.

In 1948, though, the discovery was around seven years old. The first rumours of *Metasequoias* living in China made their way to the outside world in 1941. The tree was, however, mis-identified as a variant of the Chinese water pine. In 1944 some branches made their way to Professor Wan Chun Cheng at Nanking Central University, who immediately eliminated the water pine ID but was unable to determine what the tree was.

On 9 January 1948 Silverman was in the office of Ralph Chaney, the professor of palaeontology at the University of California in Berkeley. Chaney opened a letter from a Chinese colleague. 'He slit open the end of the envelope, dumped the contents on his desk, glanced at them briefly and then more carefully,' said Silverman. This was the moment when dawn redwoods and the twentieth century collided.

China at the time was in a state of political and social chaos, with communism sweeping the country. The central government was weak and bandits ruled over vast areas making travel, especially in rural areas, exceedingly dangerous. Early in March 1948 Chaney, afflicted with asthma and fatigue, and Silverman, suffering at the hands of Chinese bureaucracy, arrived at the village of Mo-Tao-Chi after nearly two weeks of travel. 'At the edge of town was a towering tree, perhaps a hundred feet high and about ten feet in diameter at the base,' Dr Silverman recalled. 'It was practically bare of needles, and its long, graceful branches were outlined sharply against the sky.'

It was a dawn redwood or *Metasequoia*. Silverman and Chaney were the first westerners to recognise the tree for what it was—a living fossil. Until then Chaney only knew of the tree through fossils millions of years old. But these redwoods were well and truly entwined into the lives of the Chinese villagers. The tree which Silverman and Chaney saw had a shrine at its base. Later in the day the two Americans reached the village of Shui-Hsa-Pa and counted around 1000 dawn redwoods.

That night, as the Chinese and American conversation began to die down, Chaney motioned to Silverman to join him for the final act in his pilgrimage of discovery. By flashlight the men walked to the dawn redwood forest. Chaney halted, went to his knees, pulled an envelope from a pocket and extracted a few small sprays of what looked like redwood needles. 'What the devil are they?' Silverman asked.

'They're needles from a California redwood,' replied Chaney. 'I took them from one of my trees in my garden at Berkeley.' He took one spray and reverently placed it at the base of the dawn redwood before them.

The discovery of the dawn redwood was of immense scientific interest because it gave an insight into the evolutionary development of the Californian redwood, perhaps the most famous tree in the world. Now the scientific community wanted to know whether the Wollemi pine could be justifiably compared to these two great living fossils. How long was its prehistoric record? What evidence was there to support the discoveries of Carrick Chambers?

Just after Christmas 1994, Mike Macphail happened to be visiting the herbarium in Sydney. Macphail is a palynologist—an expert in the study of fossil pollen—based at the Australian National University in Canberra. He bumped into

Ken Hill and mentioned the Wollemi pine. Like all those working in the field of palaeobotany he was wondering where this tree had come from. His imagination had been captured by its discovery, except his mind was working at a different scale—while the rest of the world was marvelling at its prehistoric beauty he was pondering its microscopic secrets. Macphail had a hunch. He asked Hill whether he could take a look at some of the Wollemi pine's pollen on the chance that somewhere on the continent it might have left a record. Hill agreed and Macphail headed back to Canberra.

Macphail's office at the Australian National University is spartan in its decoration, rich in its superficial appearance of clutter and deceptive in the quantity of knowledge stored there. Against the wall is a filing cabinet full of index cards which half obscures a portrait of Samuel Johnson, giving visitors the impression that the great English writer and taxonomist of words is quietly spying from behind the furniture. When I asked Macphail about the picture he explained that he admired Johnson for ignoring the lure of a peerage. In the palaeobotanical world Macphail is also regarded as a researcher with a deep commitment to rigour and a healthy cynicism about the slippery poles that scientists have to climb within their institutions.

To be a good palynologist requires exceptional book-keeping and filing skills because in the course of a career a pollen expert will examine and record millions of microscopic fossil grains. This is the equivalent of sorting the grains of sand in a child's sandpit and having them on hand for instant retrieval.

Within a month Hill sent Macphail a Wollemi pollen cone. Macphail dissected it and placed a slide containing pollen beneath the lens of his photo microscope. When he examined it he saw a familiar sight—a simply formed pollen grain that consisted of a sac-like sphere, covered in minute dots. The grains were almost identical to fossil pollen grains up to 90 million years old which scientists, mostly in the oil industry, had studied for thirty years but were never able to link to something alive. What Macphail had known throughout his career as *Dilwynites* was now known around the world as a Wollemi pine. He was ecstatic. Until that moment *Dilwynites* fossils had been frustratingly omnipresent but no-one knew which, presumably long-extinct, plant had been producing them. The mystery was heightened because two to three million years ago *Dilwynites* had all but disappeared from the fossil pollen record. Macphail was inclined to believe that whatever had produced *Dilwynites* had become extinct. And yet it remained one of a small collection of fossils that he and his colleagues thought—dreamed—was still alive. To most of us two million years beggars comprehension but to a palynologist such a slab of time is a wink. Plants have a far greater capacity than animals to hide for extended periods of time and Macphail knew that, while it was apparent that whatever had produced *Dilwynites* had experienced a population crash, there was a small chance that somewhere a *Dilwynites* tree existed.

Dilwynites was first found in the 1960s by a young researcher named Wayne Harris, who was working on the Victorian coastline. In November 1963 Harris was a twenty-six-year-old graduate of the University of Adelaide. He had majored in geology and botany and had decided to pursue palynology as a career. He took a job with the South Australian Geological Survey. At this time the true richness of the oil deposits buried under Bass Strait, the shallow but wild stretch of water between mainland Australia and Tasmania, was becoming apparent. Teams of young scientists were swept up in the national adventure of putting Australia on the map as one of the great resource-rich regions on the planet. Harris's boss sent him to a tiny village in southern Victoria called Princetown situated in what is known to geologists as the Otway Basin. This basin—a vast geological depression which over millions of years has been filled with vegetation and sediment—bridges South Australia and Victoria. Harris trawled through the literature before he sampled in detail a section of sediments east of Princetown.

A lot of shelly fossils were found throughout the layers that Harris sampled, indicating that the area, when the sediments were deposited, was a shallow lake that was at times probably open to the narrow stretch of water between the new Australian continent and Antarctica. The layer of sediments Harris was interested in represented about five or six million years of Australian history. He was working in a section of soil deposited approximately 45 million years ago. Over a period of a few cloudless summer days, Harris scraped together around thirty soil samples off the cliffs. These collections consisted of both organic and mineral matter. Back in his laboratory he began the long task of processing his material. First he mixed the sediment with hydrofluoric acid which, with a deadly fizz, dissolves most of the unwanted mineral matter. What remained was selectively exposed to nitric acid, getting rid of all the soluble organic matter. And there was the botanical equivalent of the sparkling flecks of gold in a miner's pan—prehistoric but unpetrified plant matter telling the story of what kind of forest once grew in this 45-million-year-old layer of dirt by the seashore in modern Victoria.

Through his microscope Harris searched grain by grain looking for pollen from both known and extinct families of plants. He recognised and catalogued *Casuarina*, commonly referred to as she-oaks, which today make an entrancing noise as their needles catch the wind and turn it into a purring, deceptive hum. On the basis of what Harris found we know it is likely that a similar sound could have been heard around these palaeo shores. There were also eucalypts and other species related to them like the lilly pilly, producing colourful and gluey fruits high in the rainforest canopy. But the most common pollen he found on the slides was the family Proteaceae. The modern richness of this group of plants, known for their breathtakingly elegant flowers, stunned Sir Joseph Banks. When, in 1770, he sailed with Captain James Cook into a waterway south of where Sydney is now located, Cook named it Botany Bay. It was a Proteaceae, *Banksia*, that was named after Sir Joseph.

One kind of pollen appeared in all of Harris's samples but failed to match anything in the scientific literature. Harris didn't know what it was and neither did his superiors so he decided to publish an article about his find. He called the strange new pollen *Dilwynites*. His work yielded a rich harvest of other new palaeobotanical discoveries. On that first field trip to Princetown Harris found four new genera of plants and thirty-one new species, including two species of *Dilwynites*. 'Genus *Dilwynites* Harris, new genus,' he wrote in his paper, 'Pollen nonaperturate, spheroidal. Exine sculpture consisting of verrucae, granulae or spinules.' Reduced to laymen's terms the *Dilwynites* pollen was round, had no entry holes and its external surface was covered by granular bumps or spines. Harris assumed that the tree that produced *Dilwynites* was in the laurel family.

As anyone who suffers hayfever knows, pollen travels through the air in massive clouds. Sometimes it is invisible and at other times it is so thick that yellow piles, billions of microscopic grains, can build up on street corners like mounds of sulphur powder. Unused pollen drops to the ground, sinks to the bottom of lakes and oceans or simply disappears. Unwanted pollen and spores have been landing on the ground for hundreds of millions of years and—with the right conditions—are indestructible. To survive, pollen cannot be exposed to oxygen. Nor can the grains be subjected to extremes of heat and pressure. When conditions are benign, pollen grains can be preserved in astronomical numbers—millions per cubic centimetre—and among palaeobotanists palynological field work is a comparative breeze. Pollen scientists need little more than a trowel and a matchbox-sized container to collect their fossils. The pollen is, however, visible only under a microscope and extracting the grains from the dirt requires the use of hydrofluoric acid—one of the most toxic reagents known and the best for isolating pollen.

If the conditions for preservation are maintained through millennia there is no limit on how long a pollen grain can survive buried underground. Such pollen is not petrified. The carapace is still the same organic polymer that was buried up to 400 million years ago. In fact the pollen wall is probably the most resistant natural polymer on earth. This means that palynologists can look at the exact pollen grains of the first primitive vegetation that began colonising millions of square kilometres of empty earth, though the delicate DNA inside is long gone and the preserved pollen is no longer capable of reproducing. Palynologists can thus make a good fist of re-creating entire forests from the earliest days of the plant kingdom. 'These are extraordinarily stable compounds,' Macphail told me. 'In the New York water supply there are 400-million-year-old spores from the Devonian era that are pristine—they look like they have just dropped off a tree.'

Pollen is measured in micrometres or microns, millionths of a metre. The smallest pollens are about eight microns in diameter and the largest around 120. The distinctive decorations on the grain are smaller than a tenth of a millionth of a metre. Most pollen grains fit into the twenty to forty micron range. Pollen can be seen quite clearly under a 125 times magnification but palynologists tend to work on about 800. The *Dilwynites* grains Macphail studies are at magnifications of up to 2000.

Within a few minutes of my arrival in his office he began searching for the perfect *Dilwynites* slide to show me. The fossil material I saw in his microscope was a golden yellow brown—similar in colour to dark amber. Around 35 million years old, it came from a fossil site called Cethana in north-west Tasmania. The *Dilwynites* fossil evidence suggests that *Wollemia* has been developing into new species for nearly 100 million years. As far as we know only one species in the genus *Wollemia* survives today but once there were many. No-one knows the purpose of the granules on Wollemi pine pollen or for that matter why the pollen is decorated at all. But, Macphail told me, while *Dilwynites* is less spectacularly decorated, pollen ornamentation in general has proven to be a boon for scientists. 'If you could put a pollen in the Archibald you would probably win. They are interesting objects in their own right and the diversity of their shape is spectacular. They're intriguing. I could show you pollen that looks like a skull. What keeps people like myself looking down a microscope is that it is almost like a treasure hunt—you never know what you are going to find.'

Once Harris had found, identified and described *Dilwynites* other pollen experts began finding it in their samples. Soon these scientists were noticing it everywhere they sunk a shovel. The discovery of the two types of *Dilwynites*— *Dilwynites granulatus* (whose pollen was almost identical to *Wollemia*) and *Dilwynites tuberculatus*—were proof that Wollemi pines were once part of a bigger and more diverse group of trees. A typical, single male Araucariaceae cone can produce ten million grains of pollen, and hundreds of billions of grains of fossil *Dilwynites* are buried throughout the south of the globe as far apart as the Northern Territory, suburbs of

Sydney, New Zealand and Antarctica. Wollemi pines or their ancestors once covered continent-scale slabs of the southern hemisphere. But it is unlikely we will know how these extinct species of *Wollemia* looked or why *Wollemia nobilis* survived and other similar species succumbed.

Once Mike Macphail had reunited *Dilwynites* and the pollen from a living Wollemi pine he hit the phones. 'After nearly ninety million years of appearing in the fossil pollen record *Wollemia-Dilwynites* became a blind spot until someone stepped on the living beast,' he told me. As excited as a palynologist can be, Macphail first contacted a fellow Canberra pollen expert, Liz Truswell, who to the curiosity of her colleagues has an eccentric passion for combining her pollen studies with her art. He told her that he had probably found a living example of *Dilwynites*. 'The granules on the Wollemi pine pollen are not as coarse as the granules on *Dilwynites* but they're in the same box. I think we have got *Dilwynites*.'

When I visited Truswell's home on a winter's morning for breakfast three years after she had taken Macphail's phone call she showed me her most recent artwork—prints of Australian pollen grains. These grains had been magnified millions of times and then depicted floating, looking like UFOs hovering above the landscape. Truswell has acted on Macphail's prophecy of fossil pollen as an art form. She has also done much of her palynology in Antarctica and, although at present a fine arts student at the Australian National University, she is still regarded as one of the nation's best palaeobotanists. Truswell too spent much of her career wondering what kind of plant had produced *Dilwynites* pollen for tens of millions of years.

In 1986—eight years before Wollemi pines were discovered—Truswell published an intriguing scientific paper. From 50-million-year-old cores collected from the Hale River in the central Australian desert, near the home of Australia's famous Aboriginal artist Albert Namatjira, she found evidence of *Dilwynites* growing with a possible species of Venus flytrap that she called *Fischeripollis halensis*. These fossils are the oldest Venus flytrap pollen grains in the world.

The rainfall in the region at that time may have been as high as 1.8 metres per year, easily allowing rainforest species like *Wollemia* to survive. Today the Hale River is a desert landscape where only the toughest, most drought-resistant botanical diehards, like spinifex and acacia, can survive. Venus flytraps are now restricted to the south-eastern part of the United States. Although Truswell found that *Fischeripollis halensis* has minor differences with the pollen of the modern swamp-loving botanical carnivores, its discovery suggests that the ancestors of the modern plants may have lived in central Australia. The fossil pollen grains of *Fischeripollis* have also turned up in South Australia and in the Gippsland and Otway Basins in Victoria. Imagine a continent comprising a carpet of ferocious-looking Venus flytraps surrounded by rainforests with emergent Wollemi pines busting through the ancient canopy. It is also possible that the foliage gigantism documented in other fossil plants from the time may have applied to Australia's Venus flytraps—think of that, plate-sized traps!

Within months of the public announcement of the discovery of the Wollemi pines, Macphail was the lead author on a scientific paper which declared that the earliest reliable occurrences of *Dilwynites* were 91 million years ago. The *Dilwynites* genus, wrote Macphail and his colleagues, was almost certainly part of the Gondwanan flora, but its lineage may have evolved elsewhere because there is no detectable difference in the time of its first appearance between southern and northern Australia. *Dilwynites granulatus* flourished in the era which began around 65 million years ago and is recorded at sites as different as the highlands of Tasmania (more than 800 metres elevation) and a meteorite impact crater in the lowlands north of Perth, Western Australia. The youngest known fossil specimens are in two-million-year-old sediments from Bass Strait. The best non-Australian record to date is from western Antarctica, where *Dilwynites granulatus* is reported as a fossil Lauraceae pollen type in Palaeocene strata on Seymour Island.

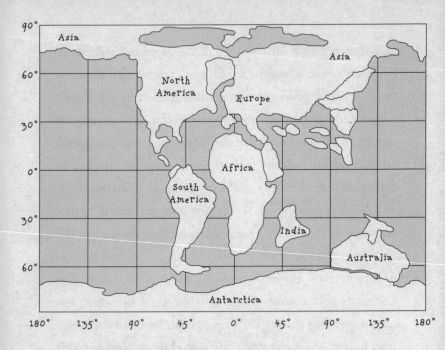

The world when Wollemia *flourished, 90 million years ago.*

The palaeobotanists also remarked that the other species of *Wollemia* that once existed—*Dilwynites tuberculatus*—had been found in Antarctica too. The sediments in which these pollens were found were badly mixed, however, and all that could be said about their age was that they were deposited sometime after the meteorite impact which wiped out the dinosaurs 65 million years ago. Macphail and his colleagues also revealed that the *Wollemia* lineage was probably present in New Zealand during the last few million years of the dominance of dinosaurs.

The Wollemi Pine

During its 90-million-year sojourn in Australia, the team of palynologists wrote, the Wollemi pine survived major warming, then cooling, of the global climate as well as the effects on regional climates of Australia's drift from high latitudes, adjacent to Antarctica, to its modern position near South-East Asia. 'Why the pine has survived,' they added, 'in apparently only one of a number of similar gorges on the eastern highlands of New South Wales is unknown.'

Dilwynites may in fact stretch back even further into the past. In 1980, Macphail told me, a 'funny thing' which could be *Dilwynites* was found in the Surat Basin, part of Australia's Great Artesian Basin. In the scientific literature it was attributed to the regal fern *Osmunda*. The fifteen-metre-deep sediments in which the pollen was found were dated at 120 million years—30 million before *Wollemia* makes a more general appearance in the fossil record.

But Macphail cautioned me about assuming that just because *Dilwynites* is the same as Wollemi pine pollen the two trees looked the same. It is a golden rule of palaeobotany to be cautious about interpreting data. The behaviour of modern trees is hard enough to understand, and habits like dormancy cannot be fossilised. It is even dangerous to assume that because pollen identical to a species alive today has been found in the fossil record it comes from a similar tree with similar habits. 'Pollens are what's known as evolutionary conservatives,' Macphail remarked. 'Other plant organs may be evolving at a faster rate. My feeling is that we wouldn't know the size of the plant that produced *Dilwynites*. It may have been a shrub. We just don't know.'

One salutary lesson for those who make the mistake of thinking that trees in the genus *Wollemia* have looked the same for 90 million years is the mystery of Huon pine. Pollen identical to that from the Huon pine is found deposited, together with *Dilwynites*, for tens of millions of years. Huon pine leaves, branches and seeds, however, do not appear in the fossil record until the last 2.6 million years. This implies that in spite of the pine's primitive appearance it may have evolved quite recently. The primitive pollen identical to modern Huon pine pollen may be from an ancestor that looked very different.

At almost the same time as Macphail was asking Ken Hill for a pollen cone from a living tree, Alan Partridge, a fifty-two-year-old Melbourne-based geologist, began asking himself the same question. Did the Wollemi pines have a fossil pollen record?

His first knowledge of the tree was from a clipping sent to him by an American palynologist colleague. 'My first reaction,' he recalled, 'was, "If that thing's up there in the mountain range it should bloody well be in my preparation. Now which pollen have we got in our assemblages that we haven't got a recent affinity for?"' He started going through his slides. 'I was terribly disappointed in one respect,' he told me, 'because when you find a new plant like that you hope it's got exciting new pollen and to me to have a drab old pollen like *Dilwynites* was a bit of a negative.'

Partridge has spent most of his adult life working for oil companies analysing drill cores, and has probably seen more fossil *Dilwynites* than any other person. He is now at the School of Earth Sciences at La Trobe University in Melbourne and has one of the biggest databases in Australia of what once grew on the continent between the age of the dinosaurs and the great ice ages that made the country a desert.

Bass Strait has proven to be the premier oil province in Australia. Its oil-producing basin was created by being filled with the sediment, trees and pollen carried by the rivers that flowed into it. Three billion barrels have been discovered in the Gippsland Basin of which nearly 2.5 billion have been retrieved. Most of Partridge's career has been spent with the petroleum giant Esso, helping the company search for pollen indicators to date the rocks. In the 1970s as a young geologist he spent time in Bass Strait on exploration rigs— ships that are basically industrial sites in the middle of the sea drilling cores twenty-four hours a day while geologists are exploring for oil. The cores are around eight centimetres in diameter and when they reach the surface it may be the first time in nearly 100 million years that they have felt fresh air. As the cores are wiped free of mud anything can appear before a scientist's eyes. 'There's a great sense of anticipation and expectation,' Partridge told me. 'It's a bit like opening a Christmas present when you are a little kid. Even though you have told Santa Claus what you want you just don't know what's in that present.'

Sometimes the cores reek of oil, at other times of gas. Often, after travelling up from depths where the pressure is the equivalent of five atmospheres, bubbles emerge from the cores as if they were full of champagne. Partridge first saw *Dilwynites* in these cores in 1968, four years after Harris published his paper on the strange new pollen. In the deepest parts of the basin the layer of *Dilwynites* pollen—mixed up with the billions of tonnes of sediment locked under the ocean floor—seems to begin at around 500 metres below the sea floor. Drilling at Anemone I rig went down 4775 metres, and *Dilwynites* was found from 500 metres below the sea floor right through to the bottom of the well. And it doesn't stop at 4.7 kilometres below the sea floor. 'In the middle of the Gippsland Basin—in the middle of the offshore area—we never get below the first appearance of *Dilwynites*,' Partridge said. This means that in the deepest part of the basin the bottom of the Wollemi pine pollen layer is beyond the reach of the drill rigs. 'When we go down between eighty-nine and ninety-one million years ago, I see about four varieties of the pollen that are subtly distinct. They are sufficiently different that you have got more than one species there, instead of having just rounded granules. Those granules have tiny spines on them. While it's a subtle distinction it's big enough from what we know in modern pollen that you can say, "That's something different." It's not going to be *Wollemia nobilis*; it's going to be *Wollemia* something else.'

Throughout the basin *Dilwynites* begins to disappear in the last ten million years. This means that the layer of sediment closer to the seabed, on average above 500 metres, reveals quantities of *Dilwynites* which further decline until the grains are absent. But, cautioned Partridge, it would be wrong to assume that *Dilwynites* suddenly vanishes altogether. The best way to describe what happens to Wollemi pines, according to him, is that there was a decline in the fossil record.

The absence of recorded *Wollemia* grains in the slides of palynologists over the last two million years marks the period when the genus underwent a catastrophic crash. Most of the remaining populations of the pines must have been wiped out at this time. At two million years ago it became unsafe for a Wollemi pine to remain prominent in the Australian landscape. The dramatic disappearance of pollen is probably proof that this was when the trees started to hide from climate change in earnest. It will be interesting for palynologists to do work on the bogs and swamps of the Wollemi wilderness and other parts of the Great Dividing Range. Until this is done the story of what happened to *Wollemia* in the great ice ages of the Pleistocene will remain a mystery.

Life will go to unimaginably dogged lengths to ensure that whatever soup of chemicals it has chosen to animate sees just one more sunrise. But very few organisms on earth can claim to have done this as stubbornly as the Wollemi pine.

It is easy to say that the pines have failed: they were once a species that dominated the southern hemisphere. But they are still here nearly 100 million years after the first *Wollemia* seedling pushed aside the mulch at the bottom of a rainforest somewhere in Gondwana and began its climb to the canopy. They may not have won the race but they have finished the course—a course which has repaid much of life's innovation with extinction.

Chapter 5

THE WOLLEMI FORESTS

I had known Professor Bob Hill for just a few hours when he passed me a shovel and began to teach me how to find fossil trees. A day later both he and I were filthy, covered in mud at a fossil site called Little Rapid River, in Tasmania's Tarkine wilderness. With my arms aching and my jeans buttered with three types of mud—wet, drying and caked—I remember deciding that he was the least academic-looking scientist I had ever met. Bob Hill is short and stocky with long blond hair tied back into an impressive ponytail. There is something of the Viking in his appearance. He is direct, down-to-earth and a patient teacher, with a skill for making complex concepts understandable. For more than a decade Hill and Mike Macphail have co-operated in their efforts to determine what Australia's vegetation was like during the last 65 million years. Hill works on the big visible plant fossils—leaves, trunks, branches and roots—and Macphail on the fossils invisible to the naked eye. Hill provides Macphail with sediment samples and Macphail tells his colleague how old the material is based on the pollen species present. Mike is also able to tell Bob what sort of vegetation grew in the vicinity of his diggings.

In 1983 Hill and Macphail proposed that Harris's eighteen-year-old assumption that *Dilwynites* pollen came from the laurel family was wrong. They argued that *Dilwynites* was in fact a monkey puzzle tree. Early in 1994, months before Noble walked into the canyon, Macphail had finalised a scientific paper on prehistoric pollen in which he had made the following tantalising entry on *Dilwynites*: '*Dilwynites* spp. Araucariaceae? Trees? Extinct?' Macphail had not ruled out the possibility that there might be a living *Dilwynites* tree somewhere in the Australian bush.

I wanted to talk to Hill to learn more about the world that *Wollemia nobilis* inhabited before it was confined to its canyon. In November 1998 I got my chance when he invited me to join him along with colleagues from the University of Tasmania and some visiting scientists from Leeds University

The Wollemi Pine

One man's mud wall is another's treasure-trove. Bob Hill has spent a significant part of his professional life extracting fossils from this road cutting in the Tarkine wilderness in north-west Tasmania.

on a week-long fossil dig in northern Tasmania.

In some places in the world fossil timber, even tens of millions of years old, refuses to be turned to stone and remains forever in the condition in which it was buried. The ten-metre-wide wall of mud-covered sediment in front of us in the Tarkine wilderness was such a place. At Little Rapid River an entire forest's worth of material had been consumed around 35 million years ago and then preserved with an incomprehensible perfection. The fossil forest debris we were striking at with our picks and pulling free was less decomposed than the timber left at the bottom of a pile of firewood over winter. Not only were pieces of wood scattered through these sediments but also millions of leaves. Some of these looked as though they had fallen only a few days earlier. Crushing another log I could smell the moulds and mildews of a rotting piece of timber. The smell was a combination of a handful of leaf litter and a dirty shower recess. 'How,' I asked Bob, 'can fossil wood smell?'

He brought the rich, brown, crushed timber to his nose. 'As the timber has gotten closer to the surface and accessible to modern fungi,' he said, 'clearly they have found something of interest in that wood, something for them to feed on.'

Continents had moved hundreds of kilometres since these trees sucked carbon dioxide from the atmosphere to manufacture their timber. Nowhere else on earth known to science, either now or in the fossil record, is there a place which is home to so many species of primitive conifers. At least twenty-six species, including *Wollemia*, have been described from this one fossil site—a freak show of biodiversity.

As I pulled at one log from an ancient Araucariaceae— possibly a *Wollemia*—a splinter drove itself into my little finger just below my nail. When I made a joke about the chance of infection from a long-dormant virus no-one laughed.

In his mid-forties, Hill is a practical and modest man

who understates his achievements. He gives the impression of being surprised by his renown as a palaeobotanist. He grew up on Kangaroo Island off the South Australian coast where he remembers beginning his passion for collecting and studying as a fossicker for shells washed up on the island's wild beaches. In 1980 when he arrived at the University of Tasmania he was a twenty-five-year-old postgraduate in the fortunate position of being a pioneer in his chosen field of palaeobotany. His scientific life has been spent exploring the thin layer of dirt wedged between the age of the dinosaurs and the age of the evolution of man. In places like Little Rapid River, Hill is dependent on fossil sites being exposed by accidents—bulldozer drivers putting roads through remote wilderness areas and cutting into hillsides, leaving a slice of world history exposed to the daylight. These gashes in the landscape illustrate how transitory life's grip is on the earth. At Little Rapid River, within a metre of the surface, the palaeobotanists are looking at soil deposited tens of millions of years ago.

An astronaut orbiting earth 40 million years ago would have seen a world vastly different from today. India, which farewelled Gondwana around 120 million years ago, was still rushing across the seas for a collision with Asia. Small islands provided stepping stones between South America and the Antarctic Peninsula. And it would be another two million years before a deep chasm between Australia and the Antarctic would become a fact of life. From outer space much of Australia would have been a carpet of green, its coastline unrecognisable because of giant marine incursions. From Uluru—the great monolith at the heart of the modern continent—it would have been a mere stroll to the beach. At that time sea levels were 150 metres above their present height. That was nothing, however, compared to sea levels between 90 and 100 million years ago—then they rose 250 metres, enough to put the summit of the dizzying arch of the Sydney Harbour Bridge under nearly 120 metres of water. No other fact that I can think of so highlights the staggering capacity of the planet to change. In March 2000 I climbed to the top of this bridge and found it impossible to comprehend what such a waterworld must have been like.

Forty million years ago sea levels again were high in a full-blown greenhouse environment, so rich in carbon dioxide that giant forests could survive months of polar dimness with barely a hint of stress. The atmosphere today contains around 360 parts per million but back then the figure was probably between 1000 and 1500 ppm. In the 35 million years since that time the greenhouse world has been disappearing. The Tarkine fossil forest that we were digging up grew in the last great period of global warmth when the world was almost universally fecund and biodiverse, when ecosystems as rich or richer than the Amazon forests were the norm.

Our present global flora and fauna is grossly diminished because we have just passed through the bottleneck of an ice age. During much of the last 100 million years rainfall was more than 2.5 metres per year, but once the ocean grew between Australia and Antarctica 38 million years ago, the global fridge door began to close. A current of freezing water started to circulate westwards around the Southern Ocean, blocking the passage of warm equatorial waters to the Antarctic and sentencing the South Pole to an ice cap and Australia to desert. The greenhouse of 35 million years ago was to become an icehouse. The buried forest we were excavating was a witness to the beginning of the hoary frosts, ice caps and desertification. This period was the beginning of the end for the Wollemi pines and standing in that preserved forest I could almost sense the trees' presence. Since then *Wollemia* has been searching for the last places on earth with a climate resembling what its ecosystem must have been like at Antarctic latitudes in times of soaking global humidity.

One of Bob Hill's British colleagues, Dr Jane Francis, accompanied us on the field trip. She has a calm, slightly stern manner but always gives the impression of being able to laugh at a moment's notice. She is also one of the world's leading Antarctic palaeobotanists though she insists on being referred to as a geologist. On several occasions during our road trip into prehistory Francis and Hill had discussed one of the thorniest issues of their profession: when and how did the most cataclysmic extinction event in history take place—the complete annihilation of woody plant life on Antarctica?

Great conifer forests once covered the lands at the bottom of the planet—almost down to the South Pole. Fossil trunks and timber have been found by scientists, including Francis, indicating that these trees grew tall and strong and in spite of winter darkness each year they were able to pack on a new ring of growth. Fossilised sections of these trees are up to seven metres in length and grew alongside grazing dinosaurs, similar to those at Koonwarra, with eyes adapted for winter darkness. As little as five metres separated the trees, indicating that even in the low-light conditions that prevailed these organisms were coping well. There is no doubt that these forests were on a similar scale to, if not grander than, any forest growing today.

Francis has spent much of her career searching for evidence of these warmer times at both of the poles. She has walked across places in the Arctic where 50-million-year-old dawn redwood needle fossils are exposed in such great numbers and in such perfect condition that they crunched beneath her feet—like strolling through a park in autumn. Near the South Pole, in the Transantarctic Mountains, Francis has hiked through huge fossil forests, carefully weaving between petrified tree stumps. She is also one of the few

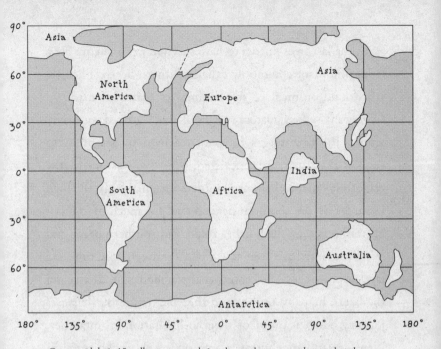

Continental drift 45 million years ago, before the greenhouse was to become the icehouse.

human beings to ever see fossilised timber collected from the Sirius Formation, located in the Transantarctic Mountains, near the Beardmore Glacier. This formation of peaks is named after the mountain on which the fossil deposit is located—2300-metre-high Mount Sirius—in one of Antarctica's relatively rare ice-free areas. The fossils of the Sirius Formation tell a stupendous survival story, culminating in a bitter endgame. Hill and Francis believe that the last forests in the Antarctic, where plants first grew 365 million years ago, were destroyed comparatively recently.

The last trees known to have survived in Antarctica are a flowering plant commonly called southern beech. They are still found on most of the landmasses that once made up Gondwana—the signature species of that outsized continent, among the first pieces of evidence used by scientists to suggest that such a place existed. This tree was one of the mightiest botanical empire-builders imaginable. In the high latitudes of the southern hemisphere, for much of the last 60 million years, the beeches ruled. But southern beech and Wollemi pines have been co-travellers through both time and landscapes—they grew together throughout Gondwana and the beech helps to understand the epic nature of *Wollemia*'s journey. There is proof of their co-habitation in Antarctica, New Zealand and Australia and, once a thorough search is complete, evidence will probably emerge that the pair grew together in South America. The fossil pollen of both species is found buried in the same drill cores through tens of millions of years of the southern hemisphere's history. It seemed only appropriate that a southern beech forest was growing on top of the fossil forest at Little Rapid River.

Southern beech is an amazingly tough tree. I have seen specimens towering in magnificent rainforests beside the lowland rivers of Tasmania and under the glaciers at the southern extreme of Patagonia in Argentina. On these Andean mountains the southern beech has views over hundreds of kilometres of glaciated wilderness, only marginally more hospitable than Antarctica. I have also seen wizened, ancient beech on the summits of frigid Tasmanian mountains, shrunken cripples so tortured by the elements that it is difficult to believe that their lucky brethren in the rainforest valleys below are the same species. In April 2000 Dr Mark Mabin, an earth scientist present when these southern beech fossils were found at Sirius, showed me some of the preserved wood in his office in Townsville, Queensland. So well had these fragments travelled through time that when they were found the scientists used a few small samples to start a fire.

After digging for the day at Little Rapid River the palaeobotanists and I retreated to our showers at a nearby motel to wash away the ancient mud packs caking our bodies before gathering for an extremely carnivorous pub feast. The Stanley Hotel in north-west Tasmania is wedged between the Tarkine wilderness and the ocean wilderness of Bass Strait. It is a place that I had wanted to visit for a long time to search for a photograph that I had been told hung in the bar. The picture was reportedly of the biggest freshwater lobster—its scientific name is *Astacopsis gouldii*—that was ever caught and which once created awe among the pub's drinkers. The biggest *Astacopsis* that I ever saw alive was a mere seventy-five centimetres long—a tiddler compared to the monsters that were caught last century. They are yabbies—the same kind of creatures hauled in from Australian farm dams using putrid chunks of meat tied to a length of string. But *Astacopsis* is the king of all yabbies and the photograph at the Stanley Hotel was of an organism of gigantic proportions. A timber-cutter held his lobster at shoulder level by its onyx-black claws, bigger in bulk than the man's forearms. Its tail hung down to his feet. A yabby more than 1.5 metres long would be one of the biggest freshwater crustaceans that has ever lived on earth. But when I arrived at the Stanley Hotel for dinner with Hill and Francis and the other scientists, I was told that the photograph had long vanished.

Over dinner our conversation returned to ancient Australia's botanical clothes and the stripping of Antarctica's. I asked whether these last trees inhabiting the Antarctic were like knee-high bonsais?

Francis looked up from her meal of Tasmanian beef. 'Ankle high, mate,' she said.

At the time of its death around three million years ago this elfin-sized forest of beech trees in the Transantarctic Mountains was so tested by the frigidity of its ecosystem that 100-year-old trees were hardly more than ten centimetres tall. These trees continued to produce pollen but sexual reproduction was so futile that the manufacture of seed ceased. What must it have been like to be the very last tree trying to survive there? The discovery of the beech trees in the Transantarctic

Mountains is like finding a single piece from a jigsaw puzzle, knowing that most of the missing pieces will never be found. Antarctica's fossil record is buried under kilometres of ice or ravaged by the movements of glaciers. For this reason no-one knows or is likely to ever know when Wollemi pines stopped growing in the Antarctic. It is also difficult to understand why southern beech has survived throughout other areas, including mainland Australia where Wollemi pines are now extinct, and why *Wollemia* has survived in Wollemi National Park while the closest stand of beech is now a hundred kilometres to the north.

Lush tall conifers in the Antarctic take some serious imagining. In February 1997 I was with geologist Greg Mortimer, a member of the first Australian team to ascend Mount Everest, and now a leader of expeditions to the Antarctic. Sitting in our ship's dining room during our voyage I asked him what he thought about continental drift. How did he allow his mind to visualise the tearing apart of continents like Antarctica and South America and the creation of oceans like the one we were sailing across?

'By thinking about it slowly,' he replied.

A few days after arriving on the Antarctic Peninsula we were pressing bootprints into knee-deep snow near the summit of an unnamed, unclimbed mountain above the L'Emaire Channel. My legs were aching, I was breathless and, despite being connected by rope to Mortimer as a precaution against crevasses or a tumble down the slope, a knot of fear was tangled in my stomach. We crossed several avalanche fields as we closed in on the ridge leading to the summit. A few days earlier as we crossed a bay in a Zodiac, and humpback whales drifted like apparitions a metre or so below the inflatable, an avalanche hit a nearby ridge. It started as a cloud that rolled down the mountain and only because it was far away did it look as harmless as chalk dust. But it was the thunderous roar which accompanied the collapsing mountain that awed us all. We staggered to the summit of our nameless peak. I knew what kind of landscape Wollemi pines need to survive, the humidity they need to stay verdant and healthy and it was impossible for me to imagine that such conditions ever existed anywhere within the thousands of square kilometres of white wilderness laid out in front of me. We took photos to prove that we had reached the crusty summit and then we sat and took in the view. Cold had obliterated all life there. All I could see was snow and ice stretching down to the gun-metal grey of the Southern Ocean.

When I was digging with Bob Hill at Little Rapid River, the presence of a modern forest growing on top of the fossil trees helped me to visualise a lush Gondwanan Wollemi pine grove. But on that lonely summit, which was unlikely to be visited by people again, it was impossible to imagine a Wollemi pine growing there, in the past or the future.

'If you want to see how Gondwana might have looked,' Bob Hill said to me, 'go to the top of any high mountain in western Tasmania—the only difference is that there are fewer species of conifer now.' We were on the second-last day of our field trip, on top of Mount Read, one of the most biodiverse mountains in Tasmania. In 1995 Mount Read made international headlines when a stunningly strange tree was discovered there, a Huon pine that was spread out over 2.5 hectares—the size of a city block. According to one estimate it was 10,500 years old, as old as human civilisation. This individual behemoth on Mount Read was already old when Tasmanian Aborigines were sealed off from the rest of Australia, around 10,000 years ago, after the glaciers melted at the end of the last ice age. At the very peak of the mountain, too, were dwarf versions of trees that were the same species as in the lowlands far below—tiny tortured individuals living at the absolute limit of their habitat requirements.

Hill talked to me that afternoon about one of the most crucial factors for an understanding of why Wollemi pines and so much of the rest of the Gondwanan rainforest species are now gone or drastically reduced in their range. 'If you increase the rainfall and the carbon dioxide as happened forty million years ago then you will get ecological compression,' he told me.

Ecological compression describes a super-abundance of the requirements for life. In the case of plants, with more carbon dioxide and water and smaller extremes of temperature, then the kinds of things that will be found living together will be more diverse. 'You could have had Wollemi pines living with everything found today on top of Mount Read,' Hill remarked as we walked to the wind-papered peak. 'By being down in the bottom of a canyon in the rainforest behind Sydney, the Wollemi pine is now hunting for an extinct climate. Australia has gone from having a relatively uniform climate and it's become much more a land of extremes. Dryness certainly took the Wollemi pines out.'

While the gap between Antarctica and its continental breakaways was increasing, and while the world was refrigerating, plants in Australia that were living on barren ridges or in extremely harsh coastal environments were beginning to experience conditions that would allow them to move into habitats that had become too fierce for trees like the Wollemi pine. This process is called pre-adaptation.

A sepia-coloured dusk was closing in on the Tarkine, bringing with it a horizon-to-horizon stillness. It was as if the wilderness rolled out all around the base of Mount Read was bedding itself down. That night there was no howling wind or pelting rain, no freezing cold to torment the relict, wild bonsai forest on the summit. Amid the knee-high canopies, with only a few cockatoos screeching in the valley below, everything was expectantly motionless. Everything living there has learned to cope with the worst. Good times, like these, are a major exception to life on the summit of a Tasmanian mountain. 'It is so quiet here,' Jane Francis said. 'Everything seems to be waiting for something to happen.'

The morning after our visit to Mount Read, the palaeobotanists and I gathered for breakfast at a cafe in Strahan overlooking Regatta Point—now a popular tourist destination for visitors wanting to explore an island in Macquarie Harbour where convicts were ferociously mistreated. The harbour is also where the Gordon River—saved from being dammed in the early 1980s—ends its journey. We ordered cheese and bacon quiches and, following Bob Hill's lead, a custard tart each. All of the fossil sites we had visited that week were remote, but that morning, Hill declared, we would be working in one of the most accessible and scenic sites in Tasmania. Pointing across an embayment in Macquarie Harbour he promised us that in no time at all we would be digging up plant fossils deposited a few million years after the cataclysmic event that obliterated the dinosaurs.

As promised, moments after breakfast Bob was sinking a pick into the dirt at Regatta Point. Tourist boats, heading out for a visit to the World Heritage area, cruised past in the clear crisp sunshine. Bob split open a sod of soil, revealing a finger-sized fragment of what had been a much bigger leaf. This leaf was more than 50 million years old but, like the timber we had found at other fossil sites, was still pliable and bendy—similar in condition to a leaf that has spent a few weeks in a compost heap. Once again, from *Dilwynites* pollen evidence, scientists know that *Wollemia* grew at Regatta Point. The fossil leaf that Bob was holding was a fragment of a much larger piece that belonged to a tree in the family Araucariaceae. When Bob and his colleagues first found such fossils at Regatta Point in the early 1980s they could not fit them into either *Agathis* or *Araucaria*.

'This is *Araucarioides*,' Bob commented, explaining that oides means 'looks like'. It was a name pulled from the hat because the Regatta Point leaves were obviously members of the Araucariaceae but different enough from the two living genera to probably be a new genus. 'If there are any fossil leaves buried in Australia that belong to the genus *Wollemia* then *Araucarioides* may be it. It's either *Wollemia* or a fourth extinct genus—it's not *Araucaria* or *Agathis*. If someone looked at *Araucarioides* and decided that it was a species of *Wollemia* then, if we were strictly applying the naming rules, we would have to change the name of the Wollemi pine to *Araucarioides nobilis*.'

But Bob, who has personally named over 100 new species of fossils—enough to fill his own little rainforest—went on to reassure me that an exception to this rule exists. The genus names *Araucarioides* and *Araucaria* would be confusingly similar. On this basis, he said, *Wollemia*, the more striking and beautiful name, should prevail. 'Finding names for fossils becomes incredibly tedious,' Hill grumbled, tapping his foot on the ground where he had just dug up the 50-million-year-old fossil. 'When we found *Araucarioides* we weren't thinking someone would find a live one.'

Regatta Point is an interesting spot for another reason—gigantism. *Araucarioides* from this location may well have been enormous. Their leaves measured more than sixteen centimetres. This is more than three times bigger than the leaves of a *Wollemia* from the modern wilderness. The average length of all the different species of leaves at Regatta Point 50 million years ago was seven centimetres. Around ten million years later at Loch Aber, another Tasmanian fossil site, the average leaf size had shrunk to about four centimetres. At Little Rapid River, as the peak of full-blown greenhouse conditions began to pass at around 35 million years ago, leaves had shrunk to two and a half centimetres. By the time of the big ice ages of the Pleistocene two million years ago, when

Australia was being racked by cold and drought, average leaf size was less than a centimetre.

These ice ages signalled an end to a golden age for southern hemisphere plants that had lasted for nearly 100 million years. Just how dramatic and sudden that change was can be best understood by considering this: Wollemi pines were still growing two million years ago around rivers near Kalgoorlie in Western Australia. Today that town is the front door to the Gibson Desert. Two million years ago grasslands were appearing for the first time in Australia as major ecosystems and the climate, which had always fluctuated, was about to turn feral. People were evolving in the African savanna and a chill dry wind was beginning to bite. An ecological upheaval—the biggest in 63 million years—was reaching a climax. The world was about to freeze.

Chapter 6

THE ICE HOUSE

I was six and my grandmother had sent me down to the corner of my street in the western suburbs of Sydney to buy some bread. Our house was one of a row of fibro cottages with red-painted concrete steps and an immaculately groomed lawn of kikuyu grass that covered us in red welts when we rolled on it, leaving us itching for days. The front gate was rickety and interrupted a row of English roses which my grandma battled to keep alive over a Sydney summer. The world of David Street, Chullora, was safe and comfortable. My grade one teacher had been my mother's grade one teacher, and nothing ever happened in that street which escaped the notice of the neighbours. It is also the first place I remember experiencing Australian weather. Summer evenings stretched on for hours, and all that winter required was an extra jumper and the occasional thick yellow raincoat when a southerly buster blew in from a mysterious place my grandparents called the South Pole.

As the gate creaked closed behind me on this particular late autumn afternoon, and I waved at Mrs Lockwood next door watering her row of roses, a stiff cold breeze hit me like a wall. It was an exhilarating wind, blowing in a steady solid mass. I remember how dry it was. At first I walked into it and then I started jogging and then running until I was sprinting across the cracks in the footpath so fast that they seemed to become a blur. I could feel the wind pulling my body skywards. I stretched out my arms and felt the lift needed to become airborne as the cold air flowed around my wings, one hand clenched tight holding the coins for the bread. I was sure that I was about to fly when I realised with a jolt that I was at the end of my runway and nearing an intersection which I was forbidden from crossing. I stopped and will never forget the sensation that if it wasn't for a few more feet of concrete I would have soared across the suburbs of Sydney.

I know now that the wind that wanted to carry me away

is a special wind. It is what has made Australia the land of gum trees, kangaroos, kookaburras and roaring bushfires, ringed around a red ocean of desert dunes. David Street faces due west and the wind that I felt was a south-westerly—the dominant force on our weather, a messenger from the freezing Antarctic ice sheet. In the last few million years it has blown away close to an entire continent's worth of vegetation. And it has done this more than once—as if the billions upon billions of tonnes of forest that it has destroyed were mere chalk on a school desk. This westerly wind is one of the signatures of the geological era in which we live. The last two million years have been a time of ice ages, so consuming and arid that, when the westerly is at its strongest, as little as 10 per cent of the Australian continent has been able to sustain woody vegetation. Until the worst of the climate changes began to grip the world from 24 million to five million years ago, Araucariaceae still grew in the inland of Australia—the Murray Basin was covered with these dry rainforests. In fact the Wollemi pine seems to have responded well in the past to limited amounts of drying, displaying an ability to move into areas from which wetter rainforests were forced to flee.

How was the continent transformed from a place whose centre once supported freshwater dolphins in vast inland lakes to a desert with dunes rolling across 40 per cent of its landmass, consuming forests as they move?

The answer is both simple and complex. The broad-brush story is that we are still experiencing the consequences of the break-up of Gondwana—a massive seaway, the Southern Ocean, now rings the bottom of the planet. The currents which flow westwards around this ocean create the Coriolus winds. These winds block warm air from entering the region and prevent cold air from being distributed more equitably around the globe. This means that cold has been locked into the Antarctic and it has become increasingly frozen—always more than ten degrees colder than the North Pole—and a bitter influence on the southern hemisphere's weather. Once a belt of unbroken water existed between Antarctica and the rest of the landmasses in the southern hemisphere the countdown to an ice age was on. The Coriolus winds began to blow almost as soon as the last moments of the Gondwanan super-continent were over. If today South America, Australia or South Africa were to be rejoined to Antarctica, the belt of ocean would be broken and the world would begin to thaw.

The complex answer has to do with Australia's huge size, its remarkably flat topography, its highly evolved flora and fauna, which have learnt to deal with extremes of weather, and the interaction of people with the continent. Each ice age has a number of peak cold periods which is known as a 'glacial maximum'. The last glacial maximum from 18,000 to 15,000 years ago brought the antipodes' wildlife to its knees.

Windblown sand-dunes are still evident on the edge of Canberra—today these dunes are covered in vegetation but just 15,000 years ago Canberra looked like Alice Springs. Some scientists speculate that the presence of people at the time added a new wild card to which 100 million years of evolution had no answers.

For *Wollemia*, an organism evolved to enjoy more than two metres of rain per year and frost-free warmth, an environment erupting with sand, dust, snow and fire was disastrous. That the tree is still alive more than 500 kilometres south of its nearest relative is hard to reconcile with anything other than miraculous good luck. But how it escaped both the freezing and the blast furnace of the ice ages will always remain a puzzle. The pine's canyon in the Wollemi wilderness was not ready for tenants until around 40 million years ago. Prior to that the entire area was subjected to a massive lava flow. It was not until the canyon's tiny creek had eroded through the basalt cap and then down into the much softer sandstones that a gorge existed where the pines are today.

So how did *Wollemia* end up hunkering down in a deep canyon? What did it drink for the thousands of years when rainfall was half what it is today? What impact did a drop in carbon dioxide levels have—from perhaps 1500 parts per million 35 million years ago to 180 parts per million during especially cold times in the last two million years? Where else was the tree hiding when glaciers covered Australian peaks and grasslands tumbled down to the coast? Did the fires of the first Aborigines, around 60,000 years ago, wipe out other populations of the trees? Did early European settlers, who cleared escarpments and valleys of forests in the Illawarra south of Sydney and in the aptly named 'Big Scrub', also chop down remnant stands of Wollemi pines?

Geoff Hope is an expert on tropical ice. Much of his career has been spent on the highest mountains on the equator in the wild places of Papua New Guinea and Irian Jaya. His main love is the Carstensz Glacier, a dying ice flow in Irian Jaya that is melting rapidly as the world warms. Because the effects of global warming are exaggerated on summits, the last fragments of the previous ice age are literally retreating up these mountains faster than scientists like Hope are able to monitor. In time there will be nowhere higher for the Carstensz ice to hide and it will cease to exist.

Hope is professor of natural history at the Australian National University in Canberra. On the morning I spent with him in the winter of 1998 a freezing southerly had blown in from Antarctica, dumping snow on the mountains which form a semicircle around the southern outskirts of the national capital. Hope pointed to the ceiling of his office and said ominously, 'The Pleistocene is just up in the sky.' He explained that snow arrives when freezing air, high in the atmosphere, drops down to lower altitudes.

On that particular morning the freezing point had come down to 600 metres above sea level. Moisture above that altitude fell as snow. The moisture was produced by a cyclone that had sat off the New South Wales coast for days, sucking water from the ocean. Simultaneously a continent-sized wall of freezing air had travelled from the Southern Ocean, across southern Australia. When these two titanic weather systems collided a climatic drama was played out in the skies of New South Wales. In an ice age, Hope told me, the freezing air higher in the atmosphere comes down and stays down. There have been around five ice ages over the last two billion years. The biggest of these is thought to have occurred before multicellular life exploded onto the world stage more than 500 million years ago. In the last 2.6 million years the world has been in the grip of an ice age but the intensity of this general cold period has been broken by spells of comparative warmth. During one such interglacial thawing 129,000 years ago coral reefs were forming as far south as Newcastle, about 130 kilometres north of Sydney. Scientists estimate that about twenty-six mini ice ages have taken place in the Pleistocene. 'In terms of biological effects the mini ice age that ended around 10,000 years ago was a doozey,' said Hope.

'This one seems to have had the greatest treelessness that we can find in the record.' He agrees with the theory that maybe people made the impact of the last mini ice age in Australia greater than those experienced previously.

Between the invading sand-dunes and the eastern edge of the Australian continent was a few tens of kilometres protected by the Great Dividing Range. If the fortress of the Great Divide's escarpment had been breached then the consequences for Australian life forms would have been terrible. 'In the last glacial maximum in Australia the freezing point came down as much as a kilometre and stayed down,' said Hope. The temperature of the Southern Ocean dropped around five degrees Celsius. The rain shadows behind mountains grew larger and as the air became more frozen the amount of water that it could hold decreased. At 30 degrees Celsius, nearly a third of the air is moisture. By the time the air reaches the kinds of temperatures experienced in the Antarctic, the moisture content is almost zero. That is why Antarctica is the driest continent on earth.

During this mini ice age average temperatures fell by nearly seven degrees. There was half the rainfall and twice the wind. The firestorm conditions which brought walls of flame into Sydney's suburbs at the beginning of 1994 would have been a regular summer demon for thousands of years during the last glacial maximum.

In *After the Greening: The Browning of Australia*, Mary White described the geography of this frozen world. 'Globally sea levels were more than 100 metres lower than today and very large areas of continental shelf were exposed round continents. As an example of what this meant in terms of changed coastline, about twenty kilometres of coastal plain lay between today's coast and the ocean off Sydney, and the Queensland Plateau where the Great Barrier Reef is today was a coastal lowland.'

Imagine Sydney Harbour as a river valley and the view from Bennelong Point, the home of the Opera House, being a paddock of grassy woodland. Bondi Beach was as far from rolling surf as Sydney's western suburbs are today. Dredging work in Botany Bay (known to local Aborigines as Kamay) has turned up the remains of preserved forests. If a weather map could have been produced at this time, 15,000 years ago, the isobars in the cold fronts would have been closer because of the intensity of the desertifying westerly winds that raked across the continent. There would have been more intense low-pressure systems such as the one that hit the morning I met with Geoff Hope. The Antarctic convergence would have moved around 800 kilometres north.

The Pleistocene ice ages, Hope told me, would have favoured two things—generalisation and the ability to spread. For an organism to survive when its landscape is iced one moment and flooded the next it must be able to move and it must be able to adapt. Both eucalypts and humans were perfectly adapted to take advantage of the ice age roller-coaster ride. Both employed fire as crucial tools in their survival kit and their sudden combination in the last glacial maximum drastically upset the flora balance in Australia. People arrived from the Indonesian archipelago at least 60,000 years ago. Claims that they have been here longer—as long as 100,000 years—have been made but are subject to fierce debate. One puzzle that has not yet been solved are massive, unprecedented charcoal layers found in cores collected from off the coast of Queensland. These cores are more than 100,000 years old and they tell a Hades-like story of continental destruction—they may also offer indirect proof of the first time that humans, fire and gum trees allied themselves in Australia.

Another palynologist who has studied these cores is Professor Peter Kershaw. He believes that the cores indicate either the arrival of Aboriginal firestick farmers or an intensification of weather cycles such as El Niño, because at no other time in the last few million years has there been such a dramatic change in the structure of Australian forests. These cores signal that a boundary exists between the great stands of Australian Araucariaceae forests and the dominance of the eucalypts. Wayne Harris, the pollen expert who discovered *Dilwynite*s, also found the oldest pollen record for a eucalypt around 58 million years old. This grain was found in the Lake Eyre Basin. The oldest date for a eucalypt macrofossil—a petrified tree stump from New South Wales—is 21 million years. Nonetheless, between the extinction of the dinosaurs and 140,000 years ago there was little change in the flora of north-eastern Australia. Then a sudden and dramatic two-step decline struck the Araucariaceae forests. 'The pattern of vegetation change in relation to global glacial climatic fluctuations,' Kershaw has written, 'was broken in the middle of the last glacial [maximum] with the almost total replacement of Araucarian and *Callitris* forests by eucalypt-dominated sclerophyll vegetation.' Sclerophylly refers to the physical adaptations trees make to dry conditions—it is the botanical equivalent of sealing up moisture leaks so that water is not wasted. Eucalypts are one of the world's great sclerophyll plants.

Evidence from two crater sites on the Atherton tableland near the current Daintree rainforests indicate that Araucarian forests succumbed between 38,000 and 26,000 years ago. This was the second step that the family took towards ignominy.

In its place arrived the gum tree. As Bob Hill and his colleagues declared in their introduction to *Flora of Australia*, excellent evidence exists indicating a dramatic increase of charcoal at the same time that eucalypts became abundant.

Australia's most famous animal is the kangaroo and its most famous plant is the gum tree. The dominance of both these highly successful organisms, however, is a relatively recent phenomenon. This eucalyptus leaf left its imprint 21 million years ago.

'The current super-dominance of *Eucalyptus* in Australia may be a relatively recent phenomenon,' the scientists wrote, 'the result of the extraordinary adaptation of the genus to fire coupled with an artificial increase in ignition associated with the arrival of Aborigines.'

The drying ice age climate probably assisted this, they said, by making the vegetation more likely to burn. By the time Europeans arrived in Australia the Gondwanan remnants had been all but banished from the main stage of

The Wollemi Pine

the continent, forced into dark wet places beyond the reach of their foes.

To try and understand how fire, competition and climate change had knocked out the Wollemi pines I met again with John Benson of the Royal Botanic Gardens. 'It couldn't cope with the drying out,' he replied in answer to my question, 'but it did adapt a bit and that is why it is able to coppice.' This refers to one of the tree's stranger traits—the ability of one tree to consist of many dozens of trunks. Coppicing means that, providing the roots survive a fire or other natural disaster, the tree can resprout.

'Before the last ice age they may have been in ten gullies,' Benson speculated. 'But with the onset of the Pleistocene they became rarer and rarer, restricted to fewer and fewer of Wollemi's gullies—each ice age that comes and goes wipes out more of them. The last ice age presumably wiped them out wherever else they were in the Blue Mountains and left just one creekline of them. The weird thing is that *Wollemia* is a conifer sticking out of a flowering plant rainforest.'

Benson also suggested that the pines may have survived while the rest of the continent was ravaged by ice-age-induced drought because their roots may have access to an aquifer, an underground reservoir of water.

In North America, where the ice sheets came down, they wiped out everything. Canada has a much more depauperate flora than Australia because during the last glacial maximum everything there was covered in ice. The species that were resurrected in the north of the planet were those that could return quickly—in this case, the conifers. The greatest areas of diversity in the world are in the tropics and include Borneo and the Amazon, areas that have escaped the climatic fluctuations of other regions, allowing longer periods of time for species of plants and animals to diversify by evolution. Still, even the Amazon forests were taken to the brink by the Pleistocene—during the last ice age much of the jungle of South America was grassland.

As Australia dried and cooled, many of its most famous creatures—kangaroos, koalas, kookaburras and emus—rose to dominance. These animals had been in the ecosystem for millions of years, but as the continent dried they found themselves at an evolutionary advantage. The Pleistocence climate was so variable that even the tough *Thylacoloeo* could not cope with the combined effects of climate change and the arrival of people. Commonly known as the marsupial lion, this animal lived throughout Australia for tens of millions of years and was probably a fierce predator which weighed as much as 160 kilograms and had a jaw like bolt-cutters. It is thought that these animals were still here when Aborigines arrived.

The answer to the question of why the pines experienced an enormous crash in their population is fairly clear. The converse—why they survived—is not. The only explanation that makes sense is sheer chance. Australian prehistory appears to be littered with the occurrence of million to one events that took place only because millions of years have passed. New research is showing that, given enough time for freak events to occur and for chance to run its course, life can find itself in completely odd places. Elephants, for instance, may once have lived in Australia. The director of the Australian Museum, Mike Archer, is aware of the existence of three elephant fossils, found as far apart as Western Australia and New South Wales. For three fossils to be found suggests elephants once lived on the continent in significant numbers. Australia, however, is surrounded by a very deep trench of water that has never been less than sixty kilometres away from the closest Indonesian island with a healthy elephant population. 'Living elephants are very capable swimmers,' Archer observed in a recent scientific article, 'some having been known to snorkel and dog paddle their way across forty-five kilometres of ocean. Is it then so inconceivable that a sea-going pachyderm every once in a very rare while scared the poo out of a bug-eyed kangaroo as it hauled its super-wrinkled, barnacle-encrusted bulk onto an Australian shore?'

Fossils have been found in South America that are

identical to many eucalypts. These too may have travelled by long-distance dispersal. This, however, does push the floating elephant theory to an absolute extreme. At those distances and because of their age of 35–40 million years, it is more likely that a South American fossil eucalypt is part of a shared Gondwanan flora. But if a continent can travel across an ocean, then why not a tree?

Chapter 7

THE LIVING TREES

On 15 December 1994, the day after the *Sydney Morning Herald* announced the discovery of the Wollemi pine, I published a second story on the tree—this time about the fact that Sydney's Royal Botanic Gardens had possession of a seedling which they would study in order to learn how to propagate *Wollemia*. Jones had collected the seedling from the site.

This first living Wollemi pine to leave the canyon was shorter than a man's thumb and had a trunk thinner than a matchstick. It was whisked straight into the tightest security that botanists can muster and it would be another four years

In the wild a Wollemia *seedling has almost no chance of surviving to adulthood—its two main enemies are darkness and crowding. As part of the research effort in the canyon all seedlings located are tagged and numbered.*

before a Wollemi pine would be placed on permanent public display. During those first twelve months visitors with clearance to view the trees in cultivation were unable to enter the glasshouse in which they were growing. Instead a Wollemi pine would be fetched from its security for a brief showing as if it were some kind of family treasure. On one occasion a seedling was brought out for a photographic shoot for the *Herald*. That little tree was one of four Wollemi pines in

The Wollemi Pine

cultivation that had begun its life in the canyon. It was being carefully prepared for the camera when the stand on which it was resting collapsed. The pot crashed onto the lab floor, sending the Wollemi pine sprawling. For a few seconds there was silence as the shock registered. Then the tree was replanted. Today it is one of the healthiest and most mature of the Wollemi pines in cultivation. It has started to develop its distinctive bubbly bark, stands at more than two metres tall and, like a teenager in need of a haircut, is covered in shocks of prehistoric-looking foliage.

Jan Allen's partner Rob Smith requested that cuttings be collected from Jones for propagation at the Mount Tomah gardens. By November 1994 roots had struck from this material. But the main propagation effort began a month later at the Mount Annan Botanic Gardens from other cuttings. The program was rapidly intensified as the botanists realised they were racing against time. Until the trees were available in nurseries the wild population Noble had found—twenty-three adults, a handful of juveniles and an unknown number of seedlings—was in grave danger. Helicopter searches in nearby canyons and up and down the gorge where the pines had been found failed to locate any other stands of the trees.

Just as there are two types of scientists who study the trees—those who study their fossils and those who study living ones—there are now two types of Wollemi pine: wild and cultivated. After millions of years of solitude *Wollemia nobilis* hit one of the most significant forks in the road that it had ever experienced. It is the job of Cathy Offord, a horticultural scientist at the Mount Annan Botanic Gardens, an annexe of Sydney's Royal Botanic Gardens, to ensure that the new road is as smooth as possible. Offord grew up in western New South Wales and her parents were the kind of people whose imaginations had been captured by the discovery in 1948 of the dawn redwood. But the dawn redwood has become the bane of Cathy's life. As the scientist in charge of cultivating Wollemi pines, people are constantly asking her, 'Why is it taking so long to propagate the pines compared to the redwoods?'

Dawn redwoods are now grown everywhere—within months of their discovery botanic gardens around the world were sent seeds and seedlings. In fact, even before Chaney and Silverman tracked down the trees in China, Chinese botanists had in 1947 already collected seeds and distributed them to colleagues overseas for propagation. But by the time of Chaney and Silverman's expedition in 1948 these seedlings were not big enough for scientists to realise their significance. After his visit, Silverman returned to the US with his pockets stuffed with seeds. Chaney also took 25,000 seeds. By February of 1951 he was able to report to his counterparts in China: 'From the seeds I have brought back with me, I have propagated several thousand seedlings. These have been distributed widely over North America and some of them are successfully growing as far north as Alaska.'

The fact that it is so hard to collect Wollemi pine material means that little is known about its life cycle. But the propagation team decided that the benefits in immediately beginning a program of harvesting seeds from the trees far outweighed the risks. In the first year a mere fifty seeds were collected. In the second a more promising quantity was obtained—600.

The seed-cones of the Wollemi pine cannot be collected by climbing the trees because not only are they right up in the flimsy crowns, they are also at the very end of extremely long thin branches. Initial seed-collecting strategies considered included using trained monkeys to retrieve the cones as well as gas-filled balloons called 'dirigibles'. The only method so far devised to successfully obtain the seed-cones is a refinement of Wyn Jones's—to dangle a National Parks and Wildlife Service ranger from a cable beneath a hovering helicopter.

The first person to undertake this dubious task was Michael Sharp and perhaps no-one has suffered so much physical pain for *Wollemia*. The first time he was winched down on the end of a cable Sharp was wearing a borrowed harness. Bill Hollingsworth later described the result: 'His butt-cheeks were the size of canteloupes by the time we had finished—the harness was a bit tight.' It was an incredibly painful experience, Sharp recalled, made worse because communications were difficult—he could talk to the pilot but not to Jones, who was directing operations from the cliff using a walkie-talkie.

In September 1995 a chartered helicopter hovered with its nose pointing at the gorge wall. Sharp was offloading gear on the pilot's side of the aircraft. He was standing with both feet on the skids as the front passenger clambered out— tripping on the chopper's 'collective', which controls the pitch of the blades. It bucked a full three metres into the air and sent Sharp flying. He flew over the ledge, his fall broken by another ledge a few metres down the cliff. If his feet had been straddling the skids the helicopter's sudden lift could have split him in two. If he had not landed on the ledge the next stop was the floor of the canyon. The chopper was brought back under control and Sharp clambered up, shaken but uninjured. But it is David Crust, another NPWS ranger, who has most often been winched down among the crowns of the pines, armed with a pair of sterilised secateurs, and jiggled like a tea bag. No-one likes authorising these expeditions. If Crust were to become tangled or if there were a sudden gust of wind then the situation could turn disastrous in seconds. There is also the threat posed to the wild trees in the event of a crash. One of the residual problems with trying to collect the seed by helicopter is that its blades, with their downward thrust, shatter the cones because they are fragile when ripe. Up to 50 per cent of cones in early visits were blown open in this way. In one incident the blades of a chartered helicopter actually severed the top of a tree.

During these expeditions Crust takes about half a minute to be winched to the end of the sixty-metre cable. He has a two-way radio in his helmet and gives detailed instructions on exactly where he wants to be positioned. The chopper usually hovers downwind of the trees to minimise the exhaust blowing

David Crust stands alongside the helicopter, hovering above the gorge containing the Wollemi pines.

onto the stand. Crust wears a rucksack on his chest and, because he must remember from which tree he collected material, the inside of the pack is divided into four or five compartments. During the twenty to twenty-five minutes he spends at the end of the cable—his time is limited by the pressure that his harness applies to his crotch—he moves up and down the trees, with the helicopter rising and falling at his command. About eight Wollemi pines can be safely harvested in this way and if Crust can gather a dozen cones on these trips it is considered a success. Throughout he is blasted by downdraft and there is the constant roar of the machine above him. It is an experience that most people would find terrifying but Crust describes it as 'challenging'.

The tree in the stand that has been subjected to the most attention from seed collectors was named by Wyn Jones and Jan Allen as the Bill Tree—after Bill Hollingsworth, the popular and professional crewman in the NPWS helicopter. Hollingsworth, an American by birth, did much of the early search work with Jones. To other researchers the tree is known as Tree One or King Billy—a somewhat confusing name as there is a Tasmanian conifer species called a King Billy pine—because it is the biggest and grandest of the adult *Wollemia*. 'It is the most multi-branched tree,' John Benson told me. 'The others are basically poles, and King Billy is producing a lot of cones.'

At around forty metres tall King Billy is the tree that most prominently sticks out above the canopy of the rainforest. A tree is a platform to get chlorophyll to the light and to out-compete other trees in the vicinity. 'Sometimes a tree will never get to the canopy,' Benson observed. 'You can be looking at a tree that is maybe decades old but it may only look a few years old. When they get to the canopy as King Billy has they power on, that's when they have really got the photosynthesis going, that's when they're producing food, that's when they can put starch on their roots. That's when they can really start building bulk. Once you get above the canopy it is like going to the gym and doing weights.'

Another person who has played a key part in the story of propagating the pines is senior horticulturist Graeme Errington, who was employed at the time at the Mount Annan Botanic Gardens. He visited the trees in the wild soon after their discovery, and his description of his first encounter with *Wollemia* in February 1995 is vivid.

'Below the canopy it is quiet. The light is dim. Lichens on the trunks of coachwoods appear to glow in the failing light,' Errington wrote in his diary. 'Sassafras, coachwoods and lilly pillys support water vines with stems as thick as my forearm. Soft tree ferns and king ferns line the creek which flows through the population of Wollemi pines. Close to the trees the occasional odd-looking leaf branch lies in the water.

A diamond python, which has fallen into the crown of a tree fern, awaits the sun so it can warm its blood and seek higher ground. A break in the coachwoods reveals a single Wollemi in the middle of the creekline.'

The Mount Annan Gardens is a few kilometres off the freeway that carves through the countryside between Sydney and Melbourne, and is surrounded by the suburban sprawl of south-western Sydney. The headquarters of the effort to cultivate *Wollemia* is a line of buildings in one corner of the gardens that is off limits to the public. Pines grown at Mount Annan are all unique in their shape—some are tall and spindly, others spectacularly bushy. They grow in pots in neat rows in two separate areas. One is outside exposed to the elements, with the only protection from the weather being a huge secure steel wall and a roof of mesh and shadecloth. Others grow in humid glasshouse conditions with a suite of other endangered Australian plants. Even here, among other rare species struggling to survive in the wild, the Wollemi pines look out of place. 'It's got an antique look,' Offord told me. 'I would describe their growth, at least in cultivation, as very fast and loose. They grow quite fast and they have a very loose arrangement of dormant buds and these buds are very easily stimulated. It could be an adaptation to living in a deep, dark canyon that it is able to take advantage of any breaks in the canopy. It's almost a plastic plant. It might even be good

Cathy Offord knows the history of every Wollemi pine in her care at Mount Annan—from which tree they were propagated and whether they were grown from a seed or a cutting.

in bonsai because it's very malleable—you can change the shape of it quite easily. It's very responsive to any sort of pruning and it is very responsive to light.'

Soil tests conducted in the canyon soon after the discovery showed that it was highly acidic. Its pH value is as low as 3.5 in some places, where most soils are about pH 5. Every time you go down a notch on the pH scale it's ten times stronger and at 3.5 the soil almost has the capacity to erode anything growing in it. 'It's getting pretty close to meaning that a lot of species wouldn't survive there,' Offord said.

The attrition rate for seed in the canyon is high. Much of it is eaten on the trees by birds and by native rats and other creatures once it falls to the ground. About 20 per cent of seed is consumed by fungi. Wollemi rainforests are so sodden that at times water can be squeezed out of every living and dead thing in the base of the gorges. Many pieces of timber have bracket fungi bursting through their surfaces, looking like neatly stacked plates from a dollhouse dinner set—in every imaginable colour. Without the infinite variety of mushrooms and their relatives the canyons would long ago have been choked to the brim with unconsumed vegetation. I have camped in parts of Wollemi where there have been dozens of thumbnail-sized, brilliant red mushrooms poking through the mulch in a miniature forest of their own.

'There isn't a lot of soil there,' Offord said. 'The plants are growing into the fissures in the rock and probably tapping into a deeper alluvium.' She and her team were also surprised by the way that the pines thrive when they are removed from the canyons. They grow more strongly and far more bushy in cultivation than in the wild, probably because in captivity they are drenched with light. Once in propagation their survival rate is higher than the average native plant. Offord told me the mortality rate of the average Australian plant in a nursery is around 5 per cent but for *Wollemia* seedlings it is almost zero.

In its secret canyon, the Bill Tree, aka King Billy and Tree One, the biggest Wollemi pine on the planet, soars towards the sky. Named after helicopter crewman Bill Hollingsworth, it may be 1000 years old.

The mysterious landscape of the Wollemi wilderness does not yield its secrets easily. Its canyons, like the one where Wollemia nobilis *makes its home (inset), are deep enough to evade the impact of bushfires, droughts and humans.*

Ken Hill, senior botanist at Sydney's Royal Botanic Gardens, unveiled a living fossil when he posed in 1994 with Wollemi pine foliage and 150-million-year-old branches from Agathis jurassica, *a close relative of* Wollemia.

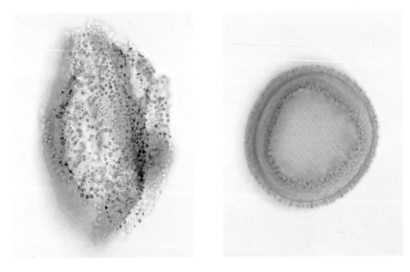

These slides convinced Mike Macphail that he had solved one of the most puzzling problems in southern hemisphere palaeobotany—which tree produced Dilwynites? *The fossilised* Dilwynites *pollen on the left, crumpled like a deflated balloon, is 17-34 million years old. The pollen on the right belongs to a Wollemi pine.*

Each Wollemi pine pollen cone (top) can release up to 10 million pollen grains to fertilise a seed-cone (right). The strange object resembling an impaled squid (above) is in fact a Wollemi pine pollen sac. Hundreds of these make up a single pollen cone.

The endgame for Antarctica's forests was played out at the Sirius Formation in the Transantarctic Mountains three million years ago. The last trees to live here were ankle-high and so stressed they had stopped producing seed.

Geologist Jane Francis, seen here in the Transantarctic Mountains beside petrified trunks from the Cretaceous period, says humans flatter themselves if they think that they can change climate as dramatically as nature does.

Defender of the wilderness Haydn Washington (above) planted an inflatable stegosaur in the branches of a Wollemi pine on the day that he discovered a second stand of the trees, and posed with his friends Christine Macmillan (left) and Bhavanna Moylan. No-one knows why Wollemi pine bark (below), with its subtle differentiation of colour, looks like so many swarming bees.

National parks ranger Michael Sharp risked his life at the end of a cable beneath a helicopter to collect Wollemi seed-cones. His harness was so tight that his colleagues still marvel that he subsequently fathered a child.

In the greenhouse Offord is running an experiment to try and duplicate conditions in the canyon. A collection of about twenty pines is being grown in 10 per cent ambient light and low pH. They are dwarfs compared to their unrestrained neighbours basking in the sunshine. The experiment shows that the canyon provides sub-optimal conditions for the trees. The Wollemi wilderness has provided the trees with a haven for millennia, but only just.

Offord and her team have gathered together as much wild reproductive material as possible, from many stages of maturity, and have learnt that the months of October and November—the southern hemisphere spring—are one of the busiest and most visually dramatic times in the Wollemi pine's year. Every *Wollemia* has both male—long thin pollen cones—and female—spherical seed-cones—reproductive organs. If the Wollemi pine is similar to other members of Araucariaceae it is possible that the entire process from pollination to sexual maturity may take over half a century in the wild.

Around October the unusually large pollen cones, the size of a man's little finger, release their clouds of minute grains. The millions of particles exploding from the cone resemble a spray from a can of insecticide. If these pollen grains do not land on a female cone and fertilise a seed then they cannot produce a new tree. The odds that pollen and seed will actually meet are extremely low. It is only for a few weeks, maybe even just a few days, that the immature female cones are open and receptive to the pollen grains. The pollen is completely at the whim of breezes, which at the bottom of the canyon are either non-existent or highly unpredictable. For every pollen grain which flukes a landing upon a receptive female cone, millions will fall onto the forest floor where they will be destroyed by the soil's high acidity. Others will drop in the creek or make their way into swamps downstream. In these swamps the Wollemi grains may be buried in just the right conditions that will one day allow them to be found as fossils, in the same way that its ancestor *Dilwynites* has been found.

Offord and her colleagues estimate that the twenty-three adult Wollemi pines found by David Noble produce about 150 female cones per year and these cones set between 3000 and 4000 viable seeds (another 30,000 to 40,000 non-viable seeds are grown). Every pollen cone—if it is similar to other members of the family Araucariaceae—produces about 10 million pollen grains. This means that of the billions upon billions of pollen grains produced, a comparative handful get to fertilise a female seed.

The proportion of unviable seed has baffled Offord. A strategy that trees employ to limit further inbreeding may, however, provide the answer. Since each *Wollemia* has cones of both sexes it is likely that when a male cone releases its pollen the female cone on that tree stays closed. In an inbred population, Offord speculated, the females may have gotten out of sync with the male cones. In other words it is possible that many of the grove's male cones are releasing their pollen when the female cones on neighbouring trees are not receptive. This is the equivalent of the boys thinking the school dance was on Thursday when the girls are dressed up to dance on Friday.

Sex the Wollemi way is a patient and delicate process. No-one is sure yet of the exact process but, if it is similar to its relatives, the seed takes between sixteen and nineteen months to mature from the moment a single pollen grain is received by the exterior of a single ovule. Around eight months after pollination a sprout will have grown from the pollen grain towards the interior of the seed ovule. When the pollen sprout reaches its destination a few millimetres away, the seed is finally fertilised. About ten months after that the seeds will be ripe, dropping between January and June. At the time the seeds are ripe their cones turn from green to brown and the branches holding them often drop off. There are approximately 250 to 300 seeds packed into each cone but only 5 to 10 per cent are pollinated and fertilised and therefore viable. The remaining 90 per cent of all Wollemi pine seeds are good for nothing but decoration.

A viable seed is instantly recognisable—it looks and feels as though it has a sunflower seed enclosed inside. These then germinate in the leaf litter with two little pre-leaves, which are very common in most plants. These pre-leaves look nothing like Wollemi pine leaves but they contain chlorophyll—the chemical crucial to the energy-producing reaction called photosynthesis. The seeds can take up to a year to germinate and this has proven to be a tortuous time-trial for Offord and her team. During one visit I made to Mount Annan, Offord's colleague Patricia Meagher produced a tiny white dish containing a seed delicately resting on a bed of cotton wool, like a diamond protected by velvet. Out of the seed a two-centimetre-long lime-green root had emerged. This is the moment when a Wollemi pine is born. The two-centimetre shoot had taken a mere twenty-four hours to grow—in the context of an eighteen-month gestation it seemed to me an explosion of life.

'That's amazingly fast,' I said.

Offord and Meagher both responded instantly. 'It has sat there for six months.'

One theory being tested is that the seed requires chilling to germinate. 'Time in the fridge,' Offord says, 'can accelerate germination to less than a month.' This ensures that the seed germinates after winter when temperatures begin to rise. The downside to this climatic strategy, however, is that the seed is more likely to rot or be eaten.

A few weeks later that little newly sprouted *Wollemia*

develops a very distinctive kink in its trunk near the base—a bit like a sudden twist in the road. The kink settles into the ground as a kind of buried root. Every Wollemi pine seedling develops this kink. By the end of its first year a Wollemi pine in cultivation stands up to twenty-five centimetres tall and has as many as seventeen branches. By three years of age a cultivated Wollemi pine can be over two metres tall and have more than seventy branches.

By April 1995 Offord and her team had also succeeded in cloning the pines. This was done in two ways: the first was a traditional method of taking cuttings and growing roots on them. The second was by tissue culture. Around 500 tiny fragments of *Wollemia* were put into test tubes in special gel. Hormones were added to make various parts of the tree grow and twenty tiny Wollemi pine clones were created, though all were incapable of growing into actual trees. Tissue culture is a means to grow many trees rapidly if there is no other alternative and, because it is tissue-specific, finding the perfect recipe of growth regulators for each part of the tree— roots, trunk, etc—is time-consuming. But members of the Araucariaceae family are tricky subjects for this means of propagation and Offord decided the best way to grow trees was from cuttings. To do this vertical shoots need to be obtained to produce trees that grow properly. Offord's team have also gone a step further and propagated the side-growing

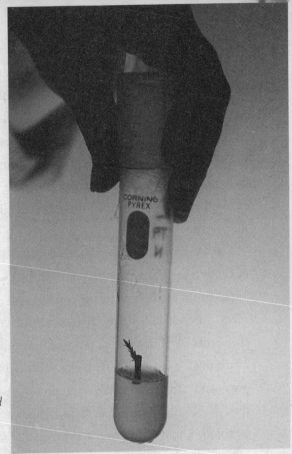

In April 1995 the prehistoric Wollemi pine was successfully propagated. This image of the 'dinosaur tree' cloned in a test tube was published on the front page of the Sydney Morning Herald.

branches of *Wollemia* to produce a conifer that grows along the ground.

In the same month the Wollemi story was to take a bizarre new twist which would have major consequences not only for Offord but for all of the scientists who had begun studying the pines.

Chapter 8

A STEGOSAUR AND *WOLLEMIA*

Haydn Washington is a bear of a man, with an unkempt beard and a voice made for telling stories. He is also the conservationist who nominated the Wollemi as wilderness. In 1974, before starting his science degree, he became one of a handful of people to walk the length of the Colo River and has since dedicated his life to protecting it. In 1979 he and other environmental activists were rewarded when 502,000 hectares of the wilderness were declared as the Wollemi National Park. A few years later Washington bought land in the north-west of the wilderness and for the last five years he has been building a castle on the highest point on his land—he prefers to call it a tower. Its doors and windows resemble Gothic arches, the whole structure is an amalgam of an Inca tower, Celtic rock fort, Scottish peel tower and a Viking fortified tower.

'Who knows, maybe it's a genetic memory,' he told me when I first saw his creation. He built the tower without a plan, on the small income he collects for his work as an environmental consultant. 'I don't have a design for a round tower made out of stone. I stuck a star picket in the ground, thought about how much was going to be involved in building various widths, drew circles on the grass, then dug a trench up to a metre and a bit deep and started building the ring foundation. The only plan I ever drew was for the rafters inside. I suppose everyone has at least one building in them.'

Getting to Washington's place entails driving out of Sydney along the giant expressway that bullies its way through the city's western suburbs. Eventually the expressway becomes the highway that has been superimposed on the route found by the early European explorers who first made their way across the mountains in 1813. After that it is a question of finding smaller and smaller roads which lead to country hamlets. From there the roads become tracks and at the end of a bumpy and muddy trail just on the edge of the wilderness is Washington's letterbox sitting on the ground beside a puddle at the entrance to his property.

Until his castle is finished he is living in a tiny two-room fibro cottage, smaller than a caravan. Here he archives twenty-five years of campaigning history. On one wall several pieces of dried Wollemi pine foliage hang above the chaos. The shack once had a bathroom but that is now filled with stores and documents. His toilet is an old dunny, fifty metres from his back door, in the middle of a disused orchard. To visit it in the evening is an astronomical experience, with the Milky Way splashed above from one horizon to the other.

The night I visited we ate quiche, drank tea boiled on a gas burner and huddled around a little wood heater as Washington recounted how he became entangled in the Wollemi pine story. He was entranced by the discovery but, in spite of his friendship with Wyn Jones, was unable to find out where the trees were located. On a weekend in the middle of January 1995 he and some friends headed off for a walk into the wilderness. They had been planning their expedition for several weeks and had decided to camp out overnight. Washington transported his team to a remote part of Wollemi in his four-wheel-drive. The walk into their chosen canyon was supremely difficult: hours were spent trying to find a route down a steep cliff, the lip of which was covered in harsh scrub that tore at the walkers' legs. Eventually they came to a twenty-metre-high dry waterfall that appeared to be impassable. Washington, however, refused to give up and after an extensive search found a navigable crack in the cliff where the wall of rock had split. The party, now battered and scratched, reached the floor of the canyon.

Washington is a quirky man with a crazy sense of humour, and he was carrying in his pack an inflatable plastic stegosaurus given to him for his birthday by Australian children's book author and illustrator Jeannie Baker. The members of the party agreed to stop every hundred metres and have a look around—they all knew that it would be easy to miss pines growing from the rock walls. The country is so broken that walkers have to stare relentlessly at their own feet.

After a while the hikers found themselves in a typical Wollemi canyon. They stopped once and then twice. Washington, who was in the lead, looked upwards and saw a gap in the canopy. 'There this thing was,' he told me. At that point they were right on the boundary between the gum trees and the rainforest. Washington had never seen such a tree in two decades of hiking in the wilderness and knew instantly it had to be a Wollemi pine. The stand it belonged to was not growing on a ledge as he had expected but on a very steep slope. As a rule, ledges in the Blue Mountains provide shelter and a diverse range of habitats for endangered plants. None of the pines was growing near the creek because it was simply too dark and narrow. The biggest tree that he saw was about twenty-five metres tall and in all he counted more than ten, though at the time he could not be sure. A couple of the trees he saw were almost like figs, growing straight out of cracks in the rock.

'These are Wollemi pines,' Washington told his astonished guests. 'It didn't look like anything else in Wollemi,' he recalled. 'It wasn't the usual rainforest plant and it wasn't sclerophyll. The others in the party looked at me as though to say, "How can you tell?"' He told them, 'This has got to be it.' They moved closer and saw the fronds and knew that what they had found was indeed *Wollemia*.

Washington then realised something very strange: there was no sign of anyone ever having been there. Not one trace.

'The structure of the trees themselves reminded me of Dr Seuss's Truffula trees because of the way the sprays of foliage come out of the ends of the branches like pompoms,' Washington told me. 'The pine is blessed or cursed, whichever way you look at it, with being a very attractive tree, with a lovely bark and a lovely form.'

The party decided to push onwards. They walked for three or four hours and were stunned to come across *another* stand of Wollemi pines, though this time there were numerous signs that people had been there including tags and seed-collection nets. They inflated the stegosaurus, propped it among the trees and took a photograph. It began to dawn on them that they might have discovered a new stand of Wollemi pines.

After the film was developed Washington sent Wyn Jones a photograph of the stegosaurus among the pines. On the back of it he wrote: 'I suppose you know about the second site several hours upstream on the tributary?'

Jones called Washington. 'Good on you. That's why I didn't tell you where the pines were because I was hoping that people like you would head out and find another site.' An indication of how well-hidden this enclave of trees is was that Noble passed right by these pines on his way to discovering the first site four months earlier.

There were seventeen trees in the second stand, which is two kilometres from the first. At first rangers and researchers from the National Parks and Wildlife Service and the Royal Botanic Gardens were elated and began to think that Wollemi pine stands might be scattered throughout the wilderness.

'I felt that I was in the presence of something very majestic and which has great power,' Haydn recalled of that extraordinary hike. 'The excitement of adding to the [pines'] population is an enduring memory.' It was nearing the witching hour inside his little cabin as he had slowly allowed his tale to unfold. It was freezing outside, the wilderness enveloped his home and the sky weighed down from above. He nestled into his chair and continued.

Washington's success at spotting the second stand resulted in an invitation to join the team studying the pines and to ride with the crew aboard the National Parks and Wildlife Service helicopter, which was conducting an aerial search for more *Wollemia*s. This was a job that nearly cost him and other Wollemi pine researchers, including Wyn Jones, their lives. 'We had been going along narrow canyons and gorges in this helicopter just brushing the sides for an hour and a half, flying around, including down main creeks and their tributaries, looking for Wollemi pines, concentrating all the time because you have only got a second and almost all the time thinking, "There!" then "Oh, it's not." We had just got out of a gorge and were starting to relax when we hit the power cable which spans the valley at that point. It runs through the air for almost a mile without any helicopter markers and so we had only a few moments. The pilot yelled, "Cable!" He yelled something else, then there was a bang and bits of perspex ricocheting around from the window. I thought we had smashed into a bird. He said, "We're going down. Crash Positions!"

'Luckily we had popped clear of the gorge and we were just above open ground. We auto-rotated down eighty to a hundred metres. I thought, "If I am going to die I would really prefer to see what's happening." So I put my head up and looked out the window. When the pilot saw the cable he didn't try to rise above it. We would be dead if he had done that. The skids would have caught it and we would have tipped over and we would all be dead. He dived down, tilted the helicopter so the rotors hit the cable and cut it cleanly. The end of the cable hit the top of the helicopter.'

The chopper came down in long grass and Washington finished the story. 'Wyn was straight-faced about the whole thing. He kept saying, "Damn, we've lost an afternoon of helicopter time looking for the pine." And I said, "Yeah, but look at it this way, Wyn, we're alive."'

Nonetheless, Washington told me, he was searching for a third stand. 'I find it hard to believe there are only two sites— but of course with any organism there must be a point where there's only three or two populations or even one that survives.' Ruminating further on why only two locations of Wollemi pines have been found he mentioned the presence of remarkably sturdy 'bullnoses' on the cliff-faces directly above both stands of trees. Elsewhere in the canyon are numerous areas clearly subject to frequent rock falls.

The next morning we hiked into a slot canyon shooting off from a gorge that Haydn had spotted from a helicopter. We squeezed into its entrance and travelled 100 metres along it. In places our outstretched arms could reach both walls of the chasm at the same time. Then we reached a slippery sheer rock wall that was impassable and, after an hour of scrambling, sliding and groping for footholds, we retreated. If there was an undiscovered grove of trees hidden behind the wall it would have to wait for another day.

Washington joined Allen and Jones on their studies in the canyon, which had been occurring since the announcement of the discovery. Watching the pair work must have been an interesting sight for Washington—Allen and Jones wore headsets with microphones so they could communicate freely in the grove. One person only was to walk among the trees relaying data to the second, standing in the creek. This method, it was hoped, would minimise damage to the trees. But by the time Washington joined them their work was being closely monitored. No-one knew anything about the trees' biological requirements. No-one understood what the pines were doing there or how they were to be managed. There is no greater proof that *Wollemia nobilis* had arrived in the modern world than to witness the scientific, legal and political problems that these trees caused.

For the first few months after the discovery Jones and Allen were given permission by the New South Wales National Parks and Wildlife Service to begin an ecological study of the pines and started to spend a great deal of time in the canyon. Just before Christmas 1995 Ken Hill and John

Benson visited the trees in the first stand. They were alarmed at what they saw and cautioned that it appeared the grove was being damaged either by unauthorised visitors or those involved in the ecological study. No-one will ever really know who was at fault in these early months but on 24 February 1995 I learned of a Royal Botanic Gardens report expressing these concerns and the *Sydney Morning Herald* published a front-page story about fears for the site. Jones and Allen began to come under the intense scrutiny of staff at both the Royal Botanic Gardens and the NPWS.

Benson noticed during his visit that the site was extremely vulnerable because it was on sandy loam soil and a steep slope. 'It wouldn't take too many people to trample that to death,' he recalled later. 'There's a lot of fibrous roots very close to the surface of the soil and we noticed that soil around two of the trees had been heavily trampled. It was obvious at the time that we needed to restrict access to the pine and that we needed to grow it in the botanic gardens. You could have a guy in there with a chainsaw and he could destroy the grove in a day.'

After this visit Benson wrote a scientific paper bluntly warning that the biggest threat the trees faced was people. A catastrophic flood, rock fall or major fire would also spell doom for the grove, he wrote. Because the odds were stacked against *Wollemia*, and its existence seemed so precarious, he also formed the view that one day people might need to intervene directly in the canyon to assist the trees. This, he said, could involve chainsawing down some of the coachwood and sassafras trees to make room for the pines. 'This would be playing God, but I don't necessarily have a problem with this if it is supported by scientific research,' Benson told me.

John Benson and Ken Hill's visit to the canyon signalled the beginning of two years of pressure on everyone involved in the project. Even the most optimistic members of the management team formed to deal with the trees knew the situation could only have a difficult ending. It did.

Throughout 1995 and 1996 Jones, Allen and Washington continued their work in spite of the Royal Botanic Gardens report. A decision by the three, taken very early after the discovery of the second site, was that this stand be left alone as much as possible, for it to be a reference for any deterioration that may occur in site one as a result of visitation. The trio visited the second site once, thereafter viewing the trees from clifftops, using binoculars. The conservation team now limited visits to site two to one a year and even then made sure that walkers remained among the stand for the absolute minimum time necessary.

The three early researchers of the trees claim that they did good and valuable science down in the gully, while both the National Parks and Wildlife Service and the Royal

Botanic Gardens felt that more control over the site was necessary. No scientific papers have yet been published on the ecological work undertaken by Jones and Allen although much of the information they gathered has been incorporated into plans for managing the trees. Perhaps conflict was inevitable: too much was at stake with this tree. The authorities had to ensure they had complete control over the site. The discoverers say they were not given enough time to complete their research, in which trunks were measured and initial studies of the canyon's ecology and the trees' life cycle were undertaken.

After a lengthy dispute Jones and Allen were given a written directive from the conservation team on 27 March 1997 to stay away from the pines. 'No site visits are to be planned or carried out until a detailed monitoring strategy has been prepared and approved,' the letter declared. 'The Conservation Team would like to take this opportunity to thank you for your field work to date in documenting the site characteristics and commencing baseline monitoring data during 1995 and 1996. This information is an invaluable reference point for further studies.' Jones, Allen and Washington have not been back to the trees since, and in July 1997 Allen complained to the press that 'it has been a very dirty fight'.

In September 1997 Haydn Washington responded to a proposed management plan for the Wollemi pines instigated by the conservation team. 'There are a number of areas that concern me,' he declared. 'The rewriting of history is one of them. Wyn Jones and Jan Allen spent two months canvassing what the pine was—a new native species—before they took their evidence to the Royal Botanic Gardens, who previously had considered a sample to be *Cephalotaxus*. Clearly both of them were involved in the botanical identification of the pine in a key fashion, yet this is not noted in the acknowledgments.'

The 27 March 1997 letter spelled the end of Jones and Allen's involvement with the wild Wollemi pines, but it was the beginning of another journey towards the understanding of the genus. After announcing the ban, the same letter continued: 'The Recovery Team has determined that one of the most essential tasks to be completed in 1997 is to finalise the patterns and extent of genetic variation between the two known pine populations. To this end, the $8000 provided by the Threatened Species Implementation account to further research on the Wollemi pine has been used to secure the services of Dr Rod Peakall, Australian National University.'

Chapter 9

Genetic scientist Rod Peakall had watched the story of the Wollemi pines unfold and knew that something strange must have happened if those trees had indeed been locked for aeons in their canyon. Most of the plants that he and his team study are refugees: their homes have been destroyed, their families have been blown apart and no-one is exactly sure what to do with them. Many are among the last few dozen of their species. In some cases only males are left and the plant can only be reproduced by cloning; others will face certain extinction no matter what happens to them. But all of Peakall's subjects have something in common—the bizarre and desperate circumstances in which they have found themselves have done weird and wonderful things to their DNA. Studying these organisms, whose lives are so abnormal, permits science insights into evolution.

The bearded, soft-spoken Peakall is a world leader in his field and it was only a matter of time before someone investigating the Wollemi pines would ask him to become involved. Wyn Jones called him first and about a month later John Benson spoke to him at a conference in Sydney in December 1995. Peakall was intrigued and leapt at Benson's invitation to analyse the *Wollemia* DNA. By the time Jones and Allen received their 1997 letters from the National Parks and Wildlife Service banning them from the site and telling them that Peakall was to be the beneficiary of the money saved, his genetic work on the tree had been underway for nearly a year. In early 1996, he had received an offer from the Royal Botanic Gardens to fund genetic work on the Wollemi pine.

Peakall had warned that the project would take time and be expensive. He needed numerous *Wollemia* samples of around five grams of fresh young leaf material. They had to be hauled out of the canyon quickly, and airfreighted overnight in ziplock bags to his lab at the Australian National University in Canberra. On some of the samples DNA extraction could be commenced immediately; others would be placed into freezer

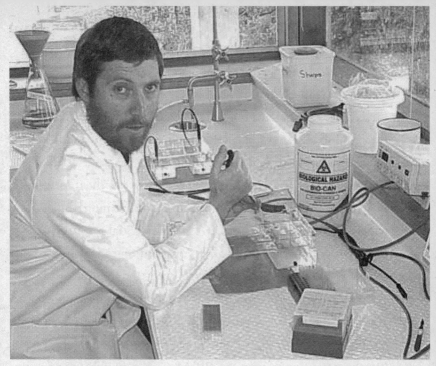

Genetic scientist Rod Peakall: he was astounded by his discoveries once he began analysing the DNA of Wollemia nobilis.

storage of minus eighty degrees Celsius. So crucial a factor was speed that Peakall requested Wollemi samples only be delivered at the beginning of the week, that way there would be no delays over weekends. It was finally agreed that for speed and security the best approach was for DNA to be extracted at Sydney's Royal Botanic Gardens and then sent to Peakall.

'This ancient lineage has presumably survived as an extremely small and isolated population(s) for many generations,' Peakall wrote in his outline of the project. 'Under these conditions, theoretical models predict a significant loss of genetic diversity over time as a result of genetic drift and inbreeding. Genetic drift is a technical term which means that if a population stays small for long enough it will lose variability until individuals almost become the same. This species may thus provide a rare opportunity to test the theory in natural populations,' he continued. 'Furthermore, it is widely assumed that genetic diversity is essential for evolutionary success, hence the goal of conservation to maximise the preservation of genetic diversity.'

On 2 April 1996 the first DNA samples from site one arrived in Peakall's laboratory. The Royal Botanic Gardens obtained the DNA by grinding down fragments of foliage. Nearly two months later, on 29 May, the first batch of Wollemi pine DNA from site two arrived. A second sample bag from site two arrived on 17 June. In all, some 100 samples were provided from the two sites and from Mount Annan seedlings. A buzz went through the university's zoology and botany departments. Peakall began to study the deepest secrets of the Wollemi pine and to this day is staggered and bewildered by what he found.

Soon after the April–June Wollemi pine DNA material had been sent, I visited Peakall to see the preliminary results of the study. He took me to a sunlit, sterile lab in one of the mazes of buildings on the ANU campus and opened the door of a freezer in which was kept a polystyrene container holding dozens of tiny tubes. At the bottom of each was DNA from one of the wild Wollemi pines. This DNA, a white glob at the bottom of each tube, contains the secrets that have kept this tree alive for over 100 million years.

The first sample that I saw was labelled 'Tree One'. This is the Bill Tree or King Billy, the biggest *Wollemia* in the stand, and it was hard to reconcile that spectacular life form with the bland little blob in the tube. Slowly, using equipment accurate to atomic levels, Peakall isolated the tree's genes. He then dissolved the DNA and, to obtain a map of the tree's genes, painstakingly ran it through the same machines used by detectives for DNA fingerprinting.

All but identical twins and clones have differences in their DNA that stand out starkly when subjected to sensitive analysis. The offspring of sexual reproduction between a pair of clones or identical twins, however, will be genetically different from the two identical parents. Genetic studies are revolutionising our understanding of life. One of the greatest scientific endeavours presently underway around the world is the mapping of the human genome, promising revolutionary changes in medical care. Every living thing has a genome, which is the total set of its genetic information—a code that governs every aspect of an organism's growth and life. In the last two decades DNA fingerprinting has transformed criminal investigation. In fact Peakall has been commissioned by the Australian Federal Police to study the DNA of cannabis. The goal of the project is to determine where seized marijuana has originated—cannabis has subtle differences in its DNA depending on where it is grown.

Genetic engineering is a technological advance that can propel life in completely new directions. DNA studies of plants have led to a total reordering of sections of the botanical kingdom and are becoming an important tool in the understanding of plant evolution. Ken Hill, for example, has used DNA analysis to work out *Wollemia's* relationship to other members of the Araucariaceae. Hill's DNA results backed up what his experience as a botanist had already told him—*Wollemia nobilis* is genetically distinct from the two other genera in the family. It seems that at some point in the distant past *Araucaria* and *Agathis* split from a common ancestor. And some time after this *Wollemia* branched off from *Agathis*.

I followed Peakall to another room off his lab, which is kept cool to protect delicate machinery, including a row of computers and the DNA fingerprinting equipment. The machine is the size of a chest of drawers and looks like something 'Star Trek' crew members would use to warp themselves to another time and place. He booted up a computer and after a few minutes a graph flashed up on a screen: the DNA analysis of a Wollemi pine. The graph displayed the results of a detailed search on part of the genome from one of the wild trees in stand one. It was generated by plotting the location of different DNA fragments. All DNA fingerprints are made up of a series of bands, which represent minute DNA fragments of varying size and intensity. Often these fingerprints resemble a barcode. On a barcode it is the position and width of the bars that permit it to be read and identified as a specific product. This is very similar to how DNA is read but when Peakall booted up the computer I didn't see a barcode, I saw a jagged line not unlike the printout from a seismograph, with the occasional major spike. A few minutes later the graph of a second tree—this time from stand two—was on the screen. When the first graph was placed on the other, the two jagged lines become one. The map of the DNA of the first Wollemi pine was indistinguishable from the second.

'When the DNA graphs from the two different trees are overlaid they are absolutely identical,' Rod told me, one hand clutching at his beard, the other resting on the keyboard as I watched over his shoulder. 'It indicates that the tree is either highly inbred or else an impossibly perfect clone. We may still find some variation but we have not been able to yet.'

So far Peakall's team has searched 1000 points on the Wollemi's genome. Although the number of points they could search is almost infinite, geneticists would normally

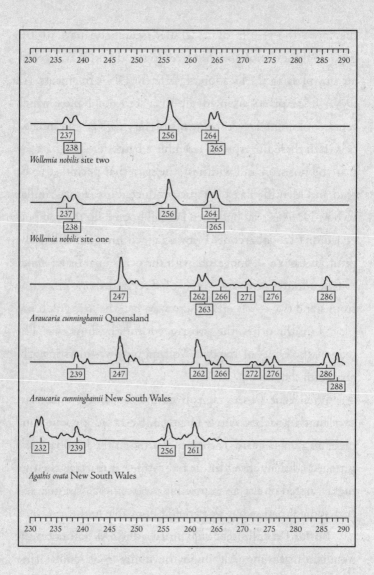

Each graph above represents the DNA profile of the same region of the genome for different members of the family Araucariaceae. The peaks and troughs represent the intensity of genetic variation. The Wollemi pine profiles are identical while the others, even among individuals in the same species, are different.

expect to find variation after searching a handful of them. This was an especially odd result because the ANU team was targeting parts of the DNA that in other plant species typically have the most variability. Peakall has compared genetic material from the crown of King Billy to the foliage from fifty seedlings produced by five different mothers. Throughout site one no trace of genetic variability was found. This is something that Peakall has never before witnessed and in fact it is unknown in the scientific literature.

Peakall's first thought was that *Wollemia* must have become a single organism. He also hypothesised that genetic drift could explain the lack of variability. 'The puzzling thing about that is that while a population of plants may become identical in theory and on computer, in the wild you can lose variability but you never have none.' Maybe, he said to me, genetic drift was being combined with clonality, that is, all of the individuals in the canyon were one big organism. To understand how both these forces can be at work at the same time, imagine a pine sending out a root that comes to the surface and then grows into a tree. When it matures the tree could conceivably reproduce with its parent. This means that both cloning and inbreeding could be at work in the canyon.

But this still did not explain why *Wollemia* had no variation. Normally a population of trees that finds itself facing extinction begins to clone itself before it loses all of its variability. This is a defence mechanism that ensures that if good times return the members of a species are not all identical. The problem became greater as soon as Peakall began to run DNA samples from site two through his equipment. 'We expected variation between the two sites,' Peakall said. A greater array of genetic material was then analysed from site two and compared to site one. Incomprehensibly, the same result was obtained. When the points on the genome were compared there was no genetic variability between site one and site two. They were totally identical.

As the results had flashed up on the computer during the months that Peakall was studying *Wollemia* DNA, a sickening feeling began to hit. In the wild and left undisturbed such low genetic diversity is not necessarily such a disaster because it appears that the tree has become perfectly adapted to its environment. But now that the wild pines are being visited diseases could be introduced. His intuition and then his machines told him that perhaps all of the trees being grown by Offord were also identical. With hundreds of genetically identical pines in cultivation a time bomb is ticking. If the population of plants in cultivation at Mount Annan and Mount Tomah Botanic Gardens—then their only locations out of the wild—were hit by disease then the entire collection could be lost. Identical genes means identical susceptibility to disease. Peakall urgently recommended that *all* pines immediately be protected by even more stringent hygiene during visits and that the cultivated populations be separated.

A fundamental tenet of evolutionary theory derives from Charles Darwin's remark in *The Origin of Species*: 'Amongst organic beings in a state of nature there is some individual variability: indeed I am not sure that this has been disputed.' Variability is the fuel of evolution, the mechanism which allows the mutations that benefit an organism to happen. If no variability can be found in a Wollemi pine then only two explanations can be put forward—evolution in the trees has ceased or it has slowed to the point where it is no longer detectable. Until the entire genome of *Wollemia* is searched this riddle remains unsolved, yet Peakall remains hopeful that variation lurks somewhere in the tree's DNA.

The easiest explanation of why the two sites are identical is that *Wollemia nobilis* is a giant clone, spread downstream from the second site and the trees are one enormous organism that has been divided in two by some catastrophe. In this theory, stems would have emerged from roots underground and become new trees. The existence of seedlings added to the confusion; Peakall was under the impression that both sites consisted only of teenage or adult trees. There was no evidence given to him that any of the seedlings were reaching a reproductive age. According to theory, if this were happening he ought not to have got the results that he did: when two genetically identical clones sexually reproduce with each other their offspring should have DNA different from its clonal parents. Even if two identical seedlings had reproduced then Peakall would expect to find variability in the population. The only other possibility was that *Wollemia* was growing its seed by apomixis—this means that the female cone is able to produce seeds without pollen. But this has never before been observed in any conifer in the world.

Even trees in far direr straits than *Wollemia* have genetic diversity—the world's entire population of *Allocasuarina portuensis* is made up of five sick-looking adult males. This tree is so little known that it does not even have a common name. It grows on the edge of Sydney Harbour and has a multi-million-dollar view straight up the embayment towards the bridge and the Opera House. The five males are all past their prime and will never reproduce sexually again. Efforts are now being made to grow the tree from seeds collected from the females before they died. This may buy some more time for the diminutive scraggly plant that also has a history stretching back to Gondwana, when flowering plants were first evolving.

Peakall's results were so odd that he requested permission to travel to the canyon in order to determine if there was something that he was missing or not understanding about the population. On 12 June 1998 Peakall arrived on site and was forced to reconsider all of his previous thoughts about the trees. The mystery of the genus deepened further. After he abseiled down the cliff, Peakall expected to see an organism similar to *Allocasuarina*—struggling to survive. But in the Wollemi wilderness he stood staring at trees that seemed to be flourishing. The vision splendid matched neither his laboratory findings nor his intuition. 'When I went into the site I thought, "This is not like other rare species which are on the brink of extinction." I think the Wollemi pine is functioning normally. It's just highly localised. There is no obvious evidence to suggest that the tree has a serious problem and, if it's been in that canyon for a long time, the opposite is the case.'

The first individual *Wollemia* he saw battered his theories even more. 'It was a strapping young teenager-juvenile probably only a hundred years old and it has come up in the wake of a tree fall,' Peakall told me. 'It made me realise that there was at least one seedling that had made it through to adulthood.' Here was the evidence that young new trees were being recruited to the adult population and that these should produce genetically distinct offspring. Clearly they were not. 'In every other respect these seem like normal rainforest trees.'

The other big shock of the day was when Peakall flew upstream to the second stand of Wollemi pines. As he hovered above the site that had been found by Haydn

Washington he realised that something extraordinary was happening in the astonishing terrain below him and that his original ideas about the Wollemi's genes needed a radical rethink. The two stands are a mere two kilometres away from each other and in the same canyon system, but for all intents and purposes they might as well be on different continents. Clearly the landscape unfolding under the chopper was so rugged it made any genetic interchange between the two colonies unlikely. Even the possibility of genetic material in the form of stems or roots floating downstream to establish the first site seemed remote.

A month later Peakall wrote to the head of the Wollemi pine conservation team, Bob Conroy. 'While the distance between the sites is not that great "as the crow flies", it now seems to me that genetic exchange among the populations via pollen flow is very unlikely given the convoluted pathway required and the density of the intervening vegetation,' Peakall told Conroy. 'For similar reasons, while it can never be ruled out, the likelihood that site one has been derived from site two also seems unlikely.'

A new theory was needed to explain Peakall's results. Wollemi pines, he told Conroy, may not have become genetically identical in the canyon. They must have been extraordinarily genetically similar before climate change forced them in there and this low variability, over thousands of years, has been further exaggerated by cloning in one big stand of pines that had somehow been split in two.

Cathy Offord's explanation on how the population had been divided was found out the hard way. Her first trip to see *Wollemia* in December 1995 was led by Jan Allen. It began to rain as the party arrived at the camp-cave Dave Noble had found on his first visit to the canyon. The Wollemi gods decided to put on a show for the scientists—a storm of the likes that maybe strikes once every few years. Wind and rain blasts pummelled the canopy and all hope of seeing the trees on the afternoon her party arrived had to be abandoned. The wilderness had made up its mind to keep these intruders where they were for the evening. The tiny creek outside the camp-cave, normally just a metre wide, surged up 1.5 metres and increased its width five times. As the stream rose the team moved further and further inside the cave. Offord's colleague, Graeme Errington, watched in awe as in the darkness tree ferns were spotlit by gashing sheets of lightning as if they were made of solid silver. Even bush rats and invertebrates sought shelter under the overhang. The cracking of log jams could be heard through the night and the party's biggest fear was that a dam would break and water would wash away their camp.

A photograph taken the next morning (below) shows Offord in her sleeping-bag, wedged into the cave, looking as if she had been in a washing machine for a month. The creek subsided as quickly as it rose. The storm, however, had answered two of her most puzzling problems—why had the trees been obliterated elsewhere, and why were the two stands now separated? 'To ask why it is not in other canyons is to discount the power of nature,' Offord told me. 'During that flood it was truly scary. We were trapped. You could get events in those canyons that wash away everything.'

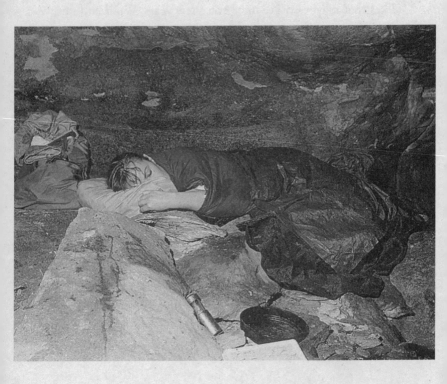

In other words where the pines weren't burnt away or hammered by rock falls, they were probably scoured out by weather conditions that are exaggerated by funnelling canyons—as if they were growing in a pot being scrubbed clean by steel wool.

Peakall sent off his report to Bob Conroy on 23 July 1998 yet he still had more questions than answers to the riddle of *Wollemia*. Soon after his visit to the pines I joined him again in his lab to discuss his thoughts. We talked about how, in the case of Wollemi pines, evolutionary theory did not sit comfortably.

'Maybe we need to think about evolutionary potential in different ways,' he said in his office, again having a good tug on his beard. He told me about some preliminary data he had obtained showing that other members of the family Araucariaceae had unusually low variability—though all except *Wollemia* had some. The monkey puzzle trees appear to be the shipwrecked sailors of the plant kingdom, finding themselves in circumstances of extreme hardship but miraculously immune to the genetic demands placed on the rest of life. Norfolk Island pines have lived for millions of years on a shard of Gondwana in the Pacific Ocean and have proven to be one of the toughest coastal conifers on earth. Other

Araucariaceae, like those on Fraser Island off the coast of Queensland, have also survived biological isolation that would have destroyed many species. 'Maybe what is happening here,' Peakall reflected, 'is that over a long evolutionary history and despite low diversity these plants have developed an all-purpose genotype.'

Perhaps, he speculated, relics like this are proof that there are other ways of surviving than by gambling on genetic diversity to ensure that certain individuals within a species do not succumb to an unexpected force that cripples other members. '*Wollemia* is likely an exception that disproves a rule,' Peakall said. 'The assumption has always been that genetic diversity is good because it is the basis of natural selection. The Wollemi pine might actually prove that in some systems it is possible to have exceptionally low variability and stay reasonably happy.'

Perhaps genetic variability is an asset to some, in particular to life's newcomers and those expanding into new ecological niches. The variability is like a high-risk investment portfolio—it increases the chance of mutations producing an ecological windfall. The downside is that it also increases the chance of a freak wipeout. Perhaps the old-timers—ancient relic tree families, which have been around since before the advent of flowers and which have experienced just about everything a plant could encounter—are able to take a different approach. To put all their assets into one account but make sure it is a safe investment. This may be why Wollemi pines are thriving and healthy but a mere two score in number.

Whatever crash-tackled the tree—one of the most conservative organisms that life has ever thrown up—must have been bordering on apocalyptic. 'So serious,' Peakall told me, 'the best genetic constitution hasn't been able to get it out of its canyon. But the flipside is, once it settled down in there, its all-purpose genome has allowed it to do as well as it can. I think there's a lot of luck in this story. Good luck in that Wollemi pines have had the constitution to hang in there in that canyon and bad luck in that whatever catastrophe drove them down there has left them stuck.'

For 100,000 millennia the atom-sized particles in a Wollemi pine's DNA have been slowly learning a little bit more about surviving. Every particle in that little white glob in the bottom of Peakall's test tubes is like a reference library containing all the knowledge and wisdom the tree has drawn on to survive.

Chapter 10

THE VISIT

Soon after dawn one day in June 1997, photographer Rick Stevens and I emerged from our forty-dollar-per-night motel rooms in a small town in the central tablelands of New South Wales. The village cannot be named for fear of suggesting the location of our destination that day—the secret canyon that is home to the Wollemi pine. We had arrived in the town after three false starts, when rain and wind had prevented any possibility of flying into the canyon. That morning did not look promising either. The horizon was glowing red as the sun rose behind dense mist and cloud—often an ominous sign of poor weather. I had never before felt so anxious about a journey as I felt about this one and the night before I hardly slept.

At a shed on the edge of town I greeted David Crust, the ranger who collected most of the Wollemi pine material while dangling on a cable below a helicopter. Crusty speaks with a laconic drawl, but his country-town manner hides a bush scientist with an understanding of the environment as deep as any professor of ecology. Wollemi is his backyard and he knows and loves the area with a passion. His region includes more than twenty national parks and nature reserves and takes in strange natural phenomena such as Burning Mountain Nature Reserve, a volcano-like burning hill that stinks of sulphur, melts shoes and sets fire to vegetation but which is a deeply buried coal seam that has been burning for a quarter of a million years.

We shook hands and Crusty introduced us to the day's other passenger, Graeme Errington, and to our pilot, Stuart Hough. The government is so sensitive about the pine that a National Parks and Wildlife Service media officer, Paul Sheridan, had been sent on the trip to monitor everything that was said by Crust and Errington. Crust produced consent forms from a manilla folder for us to sign. Fewer than twenty people knew the exact location of where we were heading. We were to promise absolute confidentiality about the Wollemi pines. 'The applicant agrees to use the phrase "in a canyon in the Wollemi National Park" when referring to the location of the site,' the consent form insisted. Rick and I had been told to ensure that everything we were wearing was newly washed, that no seeds were stuck in our shoes and that nothing was left in the cuffs or pockets of our clothes. We signed the forms.

Over a coffee, Crusty briefed Rick and me on what to expect on our flight into the canyon, telling us at the end that the helicopter would not be able to land at the top of the gorge. As it hovered we would have to leap onto a ledge halfway up a cliff. 'Jumping from a hovering helicopter is dangerous, no different from leaping from a canoe. Jump too hard and the chopper will tip,' he told us. 'When you land on the ledge, if you slip you will fall. Move away from underneath the rotors as fast as you can. Don't go down towards the tail rotor under any circumstances.'

What David Crust was telling us sounded nuts. But I had waited more than two years for a chance to see these trees and nothing was going to stop me getting on that helicopter. Rick just smiled at me. As the chopper began to whine into life in the freezing cold we climbed aboard and Crust pulled out a collection of tea towels, blindfolds for our flight to the gorge. Stevens, Sheridan and I were blindfolded. Mine was a faded gingham number, so soft and warm that it felt as though it had dried a thousand plates. Once it was secured across my face everything became a dull pinkish blur and the only thing I could see for the next twenty minutes was a faint outline of the sun. I hoped it was burning away the early morning mist.

Crust told us he would remove our blindfolds once the helicopter had entered the maze of gorges and canyons that tear through Wollemi. Within the walls of sandstone it would be impossible for us to distinguish any important landmarks and the secret of the trees' location would be safe. Sooner than expected I felt the tug of a hand on my blindfold and we were flying the thundering helicopter inside a gorge leading to the canyon hiding the pines. Crusty passed me the tea towels to look after and I shoved them under my sweater. He then gestured dramatically below the helicopter.

I saw it at once. Sticking out of the rainforest canopy, like a feathered shaft, was a Wollemi pine. Below our hovering chopper it was obvious that the tree was engaged in a fight for space. Wedged between the walls of sandstone, the vegetation looked incredibly dense and, as the helicopter continued to roar, the canopy resembled a wild green sea, whose waves and troughs had been snap-frozen. It was clearly tough for that pine tree down there—a non-flowering plant being swarmed upon by a phalanx of more recently evolved rainforest flowering plants. On the ridges, as far as the eye could see and spilling down almost to the rainforest, were the eucalypts. Fire-loving, immensely diverse and able to cope with almost anything that Australia can throw at them, gum trees own the Wollemi wilderness. In this environment *Wollemia* is like a marathon runner burdened with a bag of bricks. The clearest evidence that I could see that this was a tree with a Cretaceous heritage was its conical crown—that crown was its most striking characteristic and my strongest first impression. *Wollemia* had evolved when continental Australia was in low-light conditions at Antarctic latitudes and the trees in those times had vertical canopies that ran parallel to the trunk which maximised the amount of low-light available. Trees that have evolved more recently have canopies that are parallel to the ground, designed to catch sunlight higher in the sky. A slow-motion fight was happening beneath our helicopter between two groups of plants which didn't like each other very much.

Crust pointed to the spear-like crown of the biggest Wollemi pine below us, the oldest and tallest on the planet. I recognised it immediately as King Billy. No-one knows exactly how old King Billy is but the best estimate is 1000 years. It probably took this tree its first century of life to reach the canopy, which means that it first poked its head into full sunshine around the time the Normans conquered England in 1066. King Billy is a reclusive biological superstar, a fossil turned green before science's eyes. It was bending slightly in the dry winter's breeze—any hint that the weather was about to worsen had vanished. The giant tree was soaking up the day and I couldn't help but let out a yell.

After a few minutes of circling around King Billy the chopper began to hover perilously close to the walls of the gorge. Its rotors seemed to be just an arm's length from grinding into the sandstone. A metre below the skids of the aircraft was the ledge onto which we were to leap, and below that was a sheer 100-metre drop into what from the air looked like an otherwise inaccessible canyon. I glanced at the pilot, amazed at how transfixed he was with concentration—almost as though he was meditating. Crust's words about jumping out of boats rang through my head. The friendly easy-going Crusty was charged with getting us out of the helicopter safely and he seemed to have been replaced by a man as fixated as the pilot on the dangerous task he was performing. He gave the impression that whatever he asked us to do needed to be done immediately, without questions or conversation. He nodded at me to get out. I undid my seatbelt, opened the latch of the helicopter door, placed my feet on the skids, pretended I was in a canoe and leapt. The chopper dipped and then the instant I was in the air it began to right itself.

After feeling the thud of the sandstone ledge beneath my feet I moved to its far end as Stuart Hough worked to keep his aircraft stable while the rest of the team jumped out and our gear was thrown onto the ledge. The force of the downdraft created by the rotors was so intense that it was hard to breathe and impossible to hear. The proximity of the chopper was mesmerising and I had forgotten to secure the bottom of my sweater where the tea towels were stored. One of the blindfolds flew out, accelerating as it left my jacket, blown downwards, outwards and then upwards as if it had been picked up by a willy willy. A second and a third tea towel were sucked away before I had a chance to grab my sweater and keep the last one secure. I was obscurely gratified to find it was my old gingham mop. Hough was too busy concentrating on keeping his beast away from the walls of the cliff to notice the disaster in the making if the towels had been sucked into the chopper's engines.

Everyone was soon out and with a nod—a very cool nod—the pilot banked the aircraft sideways and downwards away from the wall and roared off up the gorge. Soon the noise from its engine was a grumble, then an echo. Within minutes we were surrounded by the silence that only those people who have ever spent time in the heart of a wilderness will experience. It was not complete—there was the sound of the wind, a yellow-tailed black cockatoo in the distance and the burble of a creek far below. We prepared our abseiling gear, shrugged on our harnesses and found anchor points for our ropes before beginning our descent. Crust chose a tree with the diameter of an arm to tie off our ropes. I was relieved he was not applying David Noble's rule of thumb. Above the gorge it was glaringly bright and warm enough to strip down to shirtsleeves, but as soon as the ropes tightened and we dropped through the canopy, gloom and damp took over—as little as a tenth of available light made its way through the trees. Underneath the forest's roof there was the cold of total stillness and high humidity that gnaws into your bones. Some of the places down in this gorge, Crust told me, get as little as one hour of daylight every twenty-four, and even without a ceiling of foliage would remain cold because the trees live at a high altitude.

Apart from a course he had taken at an indoor climbing centre at North Sydney and a few lessons with his wife, Rick Stevens had never abseiled before. He is a proud man and was determined to look like an expert. Stevens is a small, super-calm individual, who sports a salt and pepper beard. He has a huge presence—in his thirty years at the *Sydney Morning Herald* he has covered major international events, from eruptions to coups. Although we have had numerous bets about it I have never been anywhere in Australia where there is not someone who knows Rick. As his rope stretched, he leaned back and glided to the ledge far below. It was not until he reached this second ledge that he admitted to Crusty and Errington that he had never rock-climbed before in his life.

The first abseil took us halfway down the canyon wall, the second to the spongy fern-covered floor where a tiny stream flowed like liquid glass. One of the first things Wyn Jones had told me about this stream was that it was full of yabbies. Within seconds of our arrival one darted off. It had nippers painted a brilliant aqua blue, opaque tail undersides and a deep blue and green carapace. Every segment of its tail was flecked with a flash of orange and its eyes were the blackest black I have ever seen. Whichever force drove the creation of this wilderness had an intense commitment to detail. Everything in Wollemi seemed to have infinite potential for scrutiny.

This was clear water down here, filtered through unimaginably old masses of leached rainforest material. Nowhere was the tiny stream wider than a driveway or more than knee deep, yet it was the creator of this canyon nearly half a kilometre deep. The stream was testament to the patience of life and evolution. Here, days, weeks and years did not matter. This was a place to be measured in terms of millennia, a site which would noticeably change only between one ice age and the next.

I filled my drink bottle to the brim. Ancient grey vines, some as thick as tree trunks, hung everywhere, like giant gnarled ropes dumped in a tangled mess around the forest. These plaited water vines were at least as old as the European occupation of Australia and the mammoth knots they formed with each other were the size of a suburban lounge room. Some of the vines were hundreds of metres long and as solid as rocks. This, I thought, is how a spider web must look to an insect.

At the bottom of the gorge was a dense rainforest of coachwood, sassafras, lilly pilly, possumwood and tree ferns. A little further downstream were towering red gums and grey gums. This rainforest was a long and narrow strip at the base of the canyon, perhaps kilometres long but nowhere more than fifty metres wide.

Massive clumps of orchids, like bunches of bananas, clung to the wall of the cliff and others lay scattered on the ground. These had been knocked off the sandstone walls onto the floor of the forest by rock slides and temporary, flooding waterfalls. The rocks in the creek and beside it were covered in toupees of moss. So fecund was the strip beside the creek that even boulders gave the illusion of being a food source for plants. One mini-van-sized rock had several different types of fern growing on just one of its vertical faces. Its flat top was an ecosystem of its own, covered in a thick forest of ferns and countless little mosses and herbs.

Near this boulder was a tree fern so large that it was impossible to embrace it fully. Its fronds were more than two and a half metres long and stuck up at an eighty-degree angle to the trunk, which was covered in rich red-brown fur that felt like the hair between a cow's horns. Hidden behind the upright fronds were eight new unfurled fronds, waiting to reach upwards towards the tiny amount of light filtering through the canopy. They were curled like the proboscis of a butterfly or a beckoning finger.

In the crown of one tree were bird's nest ferns, each with a diameter of around two metres. On the underside of fallen trunks were dangling strands of lichens, hanging like wispy beards.

Near where our abseil had ended was the camp-cave under the overhang where so much Wollemi pine drama had been played out. I recognised the spot Cathy Offord had told me about, the place where she and her colleagues huddled when they were caught in a flash flood at the end of 1995. It was also the place that Haydn Washington called Firefly Gully because of the illuminated insects he watched there on a summer's evening. Firefly Gully is also as far as David Noble got on his first trip into the canyon and I could see why every person who visited was awed by its beauty. Tree ferns were everywhere and a waterfall tumbled over a ledge. The rainforest here was more like a subterranean cavern than a jungle. Under the canopy everything was bathed in green darkness. This was the place, Washington once told me, that evoked for him the Mirkwood that Tolkien's hobbit had to pass through on his way to the treasure in the dragon's lair.

Firefly Gully, named by Haydn Washington, is the Wollemi pine base camp—researchers retreat to this magnificent site when their day's work is completed.

We began to head upstream. Underneath another overhang, a few hundred metres from where our ropes had been left dangling, we came to a rock where park rangers had set up a primitive but supposedly effective foot bath. Here we soaked our shoes, disinfecting them against any possible pathogens which could spread to the Wollemi pines. Dieback is as deadly to a tree as the most virulent cancers are to people. The diseases are almost invariably caused by a variety of root fungi which infects the plants and deprives them of

The Wollemi Pine

food until they die. Dr Brett Summerell, a plant pathologist at the Royal Botanic Gardens in Sydney, deliberately infested, in strict quarantine, cultivated Wollemi pine seedlings to determine their susceptibility to disease. The impact of the common fungi to which they were exposed was devastating— all of the trees were dead within weeks. The canyon would be a paradise for disease, Summerell wrote in a briefing for the director of the Royal Botanic Gardens. 'The nature of the site is such that it would facilitate a high level of activity of the fungus.'

No unnecessary equipment is allowed to travel any further up the gorge. Beyond this point we were in one of the most sensitive and scientifically significant ecosystems anywhere in the world. We walked another few hundred metres upstream and then, just off to our left, we encountered our first Wollemi pine. The light began to change markedly at that moment, exactly in the way that previous visitors had described. Jan Allen's memory of her introduction to the trees captured the moment. 'The thing I noticed the most was the shift in the light,' Allen recalled. 'The shift from bluey-green that you get under the rainforest in the light because of the coachwoods and other typical species; the trunks are mostly dark coloured and mottled with lichens so it's like Prussian blue. Then, all of a sudden, it was almost imperceptible, but there was a change in the light. It changed to a very light brown or green. To see the bubbles on the bark was like when bees are swarming and they land on something and they absolutely cover it.'

The first Wollemi pine that welcomed us to the grove was a juvenile, even so it was still decades old. This was the tree that would change Rod Peakall's entire theory about the genetic puzzle posed by *Wollemia nobilis*. I was struck by how skinny it was and how prehistoric was its vegetation. I could also see why it had been so easy for Wyn Jones to mistake the tree's foliage for that of a fern. For the first time I saw the bubbly bark. It was even weirder than I could have imagined and just like Coco Pops as everyone had said.

Rick and I were both stunned into silence and we staggered on towards the grove proper. Soon Wollemi pines were all around us. I was astonished by how small and densely packed the stand is. A species which once grew throughout Gondwana is now reduced to two fragments in an area hardly bigger than a large backyard. And yet this was still a forest of Wollemi pines, not a forest with *Wollemia* in it. I found King Billy with its huge barrel trunk and ran both of my hands over its bark. The millions of bumps were surprisingly hard and cold and, every few metres, foliage was bursting out of the trunk, like a patch of hair missed during a shave. Thousands of waxy green leaves stuck out like fins from its branches, looking exactly like the tail of a stegosaurus. This was winter, a quiet time of the year, so I could not see any of the seed-cones or pollen cones.

The ground was strewn with Wollemi pine debris. There was so much material rotting on the forest floor that it felt springy. The trees' coppicing habit, the way one specimen can have dozens of trunks, means that in places the pines look like a wall of bamboo, rather than relatives of the giant Araucariaceae that line the foreshores of Sydney beaches. It was all weird, weird, weird—not a bit like being in an Australian eucalypt forest. The colour was different, the leaf litter was different, the light was a strange misty green. I spun round and round, walking from tree to tree and as I walked an odour rose from my hands. After rubbing them against the bumpy bark I had covered them in a pungent, sticky sap designed to protect the trees from the tendency to rot in such a wet environment. I understood why these trees would strug-gle in a fire-prone continent like Australia. With their high sap content I could imagine them exploding if ever a big wildfire did get down into the canyon. Crusty showed me one of the trees that had fire scars, perhaps caused by a burning tree falling from the clifftop far above. Fire is not the only thing that falls down from above—another tree is scarred by damage caused by a rock fall.

Errington and Crust set to work almost immediately. The helicopter would return in a few hours. The men had two main tasks—to collect seed that had fallen into special hi-tech nets and to check automatic data-loggers, which were recording climate information in the canyon. The seeds they were looking for were like miniature stealth bombers, with wings that allowed them to float gently from the crowns of the trees forty metres above, and up to fifty metres away from their parent. The job was time-consuming because the nets catch everything that drops from the canopy, including insects.

I walked slowly around the site while Errington and Crust worked. The strangest thing about the pines was that they actually looked primitive, as if they lacked all the fancy accessories that trees have evolved in the last 100 million years. They looked like ferns on steroids. One thing they did boast though, was individuality—as Offord had said, *Wollemia* is notable for its plasticity and it was clear that to survive down there a tree needed to be able to make abrupt changes in direction to take advantage of the light. Wollemi pines may not have any genetic variability but in appearance I had never encountered a more varied lot. Fat ones, skinny ones, bent-over ones, others growing out of the remains of dead predecessors and fig-like ones all surrounded me. Some were living on tiny 'floodplains', others on sandy ledges and others still out of cracks in the sandstone.

As they worked, Errington and Crust commented on how the canyon hides everything it contains. Even the giant seed nets were invisible until we had almost stumbled across them. After an hour or so we all stopped for lunch and sat on a great fallen trunk. This trunk was known to Jones, Allen and Washington as the Sara tree after Jones's former partner, whose strength and resilience he admired. The Sara tree is broken off close to its base. Although the trunk is long dead, the stump is a several-metre-tall mass of wooden splinters, and coppicing has taken the place of the former tree. While the biggest part of the tree is dead, its new sections are going gangbusters. Washington has formally suggested to the management team in charge of the pines that this rotting trunk, which would have been taller than King Billy, should be cored for dating.

All around us were tiny seedlings, each with two minia-ture leaves. They didn't resemble Wollemi pines. Most were little-finger height and looked as though they were just a few weeks old. In that environment it was hard to imagine a more fragile beginning to life. Some had sprouted in spots where they were obviously doomed—in the middle of streams or beneath the deep gloom of an older tree which would almost certainly kill the seedling with its shade. Others have even been found germinating among the rotting vegetation gathered in the crowns of tree ferns. Some were at risk of being smothered by the leaf litter which, Errington told me, is thirty centimetres thick in places.

Haydn Washington had also discussed the hardships

faced by little Wollemi pines. 'Some of the seeds were growing in really dry sandy environments. The thing that interested me was that we were playing God just by being there. Walking very carefully, we were inevitably opening things up and we were actually helping seedlings to germinate. They were advantaged by us being around. In some places we were having seedlings come up right under where we had cleared a bit for a net and so then we would have to move it. It is hard not to play God. One time when I went down there something had pulled out a little seedling and it was sitting on the ground. So I poked a hole in the ground and put it back in. Whether we should be doing that I don't know. But I thought it was probably pretty harmless.'

The only hope for a seedling is that early in its life a hole forms in the canopy far above. Scientists working on the site have discovered the mortality rate of the seedlings is almost 100 per cent. This is the exact opposite of what has happened with the cultivated seedlings. But on a Wollemi pine time scale such odds are fine. As long as some seedlings are given the opportunity to burst through the canopy every few hundred years then the colony will survive.

The odds of even becoming a seedling are hardly promising. When scientists found that many seeds were being washed away they speculated that this would help spread the species downstream. Noble, Jones, Allen and

Washington comprehensively searched on foot within fifteen kilometres of the two sites without finding a trace of other colonies. Researchers are now resigned to what seems to be an almost unfathomable truth—these giant trees are found nowhere else and, although they are healthy, are just one major disaster away from extinction in the wild.

The space-age nets scattered beneath the mature trees have proven to be the most efficient and environmentally sensitive way to collect seeds. It took Errington and Crust about two hours to empty all nineteen, sort through the litter that had been caught, bag everything and then label it. They worked quietly, looking every bit like gold miners sifting for treasure.

'Here's a viable one,' Errington announced, holding up a seed with a tell-tale fat middle indicating its fertility. 'We are also searching for anything else in the nets that may be of interest to scientists. We even keep the beetles, which are sent to the Australian Museum for classification.' So far the insects that have been found associated with the Wollemi pines are known species but entomologists are almost certain that living inside the trees must be some weevils and beetles that have travelled through time with *Wollemia*.

Many other organisms found living in the grove are, however, new to science—of fifty species collected among the trees around one third are new, including lichens and fungi.

Brett Summerell has also found four new potential antibiotics. He stumbled across a strange compound being produced by fungi among the pine's roots. In 1996 this was sent to Professor Gary Strobel at Montana State University. Strobel identified the compound as taxol—a powerful anti-cancer drug used against breast and ovarian tumours. The only other source of taxol is from yew trees in the northern hemisphere. It is still a mystery how this compound came to be in the canyon.

Errington and Crust downloaded, onto laptop computers, data from dozens of special sensors buried in the ground. These were monitoring variables like humidity and temperature—information that is invaluable to the horticulturists attempting to propagate the trees. To get among the pines where the data-loggers were located meant scrambling across the creek and up a waist-high bank. In places I was walking on my toes because of the large number of seedlings. Above us was a steep slope followed by a ledge on which were growing the most coppiced pines. The soil up here was almost sandy, and the trees around us looked as though they were doing it tough. It was while I was in this part of the grove, sitting watching Errington download data, that something really odd happened—a distant rumble came up the valley, getting closer and closer. It was the sound of a jumbo on approach into Sydney airport. I imagined the seatbelt-fasten sign coming on and flight attendants pacing the aisles to make sure laptops were turned off and tray tables secure. I said to Errington how incongruous the sound was and he didn't even look up. 'We hear them all the time,' he said.

With the computers full of new climate information, we collected our gear and started to head back downstream. There was one last special sight that Errington and Crust wanted me to see. It was a view from a rock up the cliff wall which is level with the spear point crown of King Billy. We climbed up the gorge on the opposite side to the rock wall we

Graeme Errington has spent years recording climate data on special equipment stationed in the grove. One of Wollemia's most striking characteristics is that a single tree can consist of dozens of coppiced trunks, making a mockery of attempts to age individuals.

had abseiled down earlier in the day. After climbing for fifty metres we emerged above the canopy and were back out into the Australia I am used to—gum trees, boronia and banksias. For the first time since leaping from the helicopter we were back into full daylight. The sun felt magnificent after a day in the darkness. The sandstone we were sitting on was warm. We clambered onto a large rock level with the crowns of the pines. We stared at King Billy and talked about our day's adventure. For me this was the moment of the day when *Wollemia* seemed most real. Then we descended again into the gloom of the Wollemi pine rainforest and began the walk back to the abseiling gear.

The last I saw of a wild *Wollemia* was a piece of bubble bark that had been washed downstream. Crusty led us back to our rendezvous point and we waited for the sound of the chopper to come roaring down the valley. When it arrived it was nearly dusk. We struggled aboard and our blindfolds were re-applied. This was a day that had passed too quickly. The chopper lowered its nose and tilted away from the canyon.

New Beginnings

After discovering the tree that honours his name David Noble completed a bachelor of applied science degree and was promoted from a field officer to a ranger in his beloved Blue Mountains. A million hectares of those mountains were listed in late 2000 on the World Heritage Register. The discovery of the Wollemi pine was central to the nomination.

Noble still heads off into the wilderness as regularly as ever and I hope the rule of thumb never lets him down. David Crust has also been promoted within the parks service but remains responsible for the protection of the Wollemi pines and the wilderness in which they live.

Wyn Jones guards his *Microstrobos*, remains actively involved in community-based environmental issues and works as a consultant. He has begun a research project studying the Aboriginal art of the Blue Mountains. Cathy Offord continues her inspiring research propagating the less fortunate members of the plant kingdom and developing new strands of popular flowering species. Offord has finished a doctorate on the genetics and reproduction of the waratah—one of the most spectacular flowers on Earth. She summed up the impact of the pines on her life: 'It has actually made things happen. It has brought conservation and biodiversity of Australian plants to the fore and that has been to me its main benefit to mankind. It has its own intrinsic value of course but, apart from filling in some of the gaps in the Gondwanan story, it has put us on the map because people can see the value of doing all this research.'

To the dismay of his colleagues Graeme Errington resigned early in 2000 from the Mount Annan Botanic Gardens and moved to the south coast of New South Wales where he has built his own home. Errington said to me before he left: 'Until now the cute and cuddlies have always got the funding ahead of any plant. There's still a huge imbalance and this story has begun to change things. Without the plants and without the habitat for plants you are wasting your time protecting the animals.'

Haydn Washington is still adding the finishing touches to his castle and also acts as an environmental consultant. He successfully fought for extensions to be added to the Wollemi wilderness and to a nearby national park, the Gardens of Stone.

Ken Hill has been unwell but remains associated with Sydney's Royal Botanic Gardens. One of his biggest dramas of the last few years was his decision to re-allocate the popular spotted gum from the genus *Eucalyptus* to the genus *Corymbia*. He says, only half jokingly, that he received death threats for the decision.

Jan Allen works at Mount Tomah and, together with the others I have just mentioned, was invited to plant a cultivated

Wollemia, to be kept inside a cage in a special grove there. Bob Hill has moved from Tasmania to the University of Adelaide and has also been appointed the head of science at the South Australian Museum. His passion for all aspects of the Australian environment, especially its flora, is undiminished. He is currently finalising a new paper that examines all the possible Araucariaceae fossils that may have been ancestors of *Wollemia*. So far the only fossils likely to be Wollemi pines or their ancestors and that are visible to the naked eye are those that were found at Koonwarra in Victoria by Carrick Chambers. This is a mystery, especially considering how widespread the *Dilwynites* pollen is. Hill is trying to unravel this botanical riddle.

Rod Peakall is in his lab in Canberra and recently had a plant he studied entered into *The Guinness Book of Records*—it is a Western Australian orchid, regarded as the best example in the world of a plant that imitates an animal. The flower of the orchid is an ingenious replica of a female wasp.

At the end of July 2000, soon after this book was first published, a third stand of Wollemi pines was discovered by a party of bushwalkers. It consists of a mere three adult trees of which the tallest is probably only fifteen metres high. The trees are around forty metres up a rock wall in a section of slot canyon 150 metres deep and no wider than thirty metres—in places it is even possible to touch both sides of the chasm. What stunned everybody involved was that the new discovery was so close to the first two stands and yet had eluded detection for nearly six years. It was in a different sub-catchment from the other two sites, which means its establishment could not have been the result of a seed being washed downstream. The third stand is most likely a remnant of a much bigger forest of pines that once filled the Wollemi canyons.

The party that discovered the third stand alarmed those involved in the management of the species. These walkers had not used foot baths or taken any of the other precautions that scientists who visit the site always insist on. Just over a year later, in September 2001, a wilderness magazine published an article in which three hikers described how they had hired a plane to fly over the wilderness in order to look for Wollemi pines. This party also walked among the trees without taking any of the necessary precautions. For the Wollemi Pine Conservation Team, preventing bushwalkers from searching for *Wollemia* is becoming increasingly difficult, and the team's worst fears for these trees are becoming real. As John Benson has warned constantly since the discovery of *Wollemia*, the greatest threat the species faces is from humans. The trees are only a few spores away from extinction due to plant pathogens imported on the soles of hikers' boots.

No-one managing the pines is sure about how to respond

to this problem. Scores of bushwalkers now know where the canyon is. Should a stand of trees be established in a wild gorge and made accessible to walkers? Should a boardwalk be built to the edge of the canyon so that visitors can view the trees from the cliff edge? Should small groups escorted by a ranger be allowed to visit one of the sites, under strict supervision? Should a sign be put in place within the grove, warning people of the threat that their presence poses? Should a foot-bath be left permanently on site? At the moment the parks service and the Royal Botanic Gardens are hoping that nothing terrible befalls the trees before a solution can be found. Until that happens, it is vital that people stay away from Wollemi pines.

Part of the answer as to why the population of trees has been splintered into separate stands comes from new research done by forest ecologist, Dr John Banks. He worked on a cross-section of timber chainsawed out of a dead pine in the first stand, whose trunk lay rotting on the forest floor. By looking at the perfectly formed annual growth rings recorded in the cross-section, Banks was able to estimate the tree was at least 400 years old. His work confirms the suspicions of scientists that King Billy is as ancient as he looks. What cannot be determined is the age of the original tree from which the rotting trunk had fallen. Wollemi pines have the capacity to re-sprout after a catastrophe—new trunks can grow from old roots that may be thousands of years old.

While studying the cross-section Banks found that in places the tree rings are horseshoe-shaped—evidence of severe wind storms. One of these enormous gales tore through the canyon in 1800 and another in 1855. The storms are recorded in the rings as clearly as a fossil is imprinted into rock. 'It would have been a wind so strong that the top third of the tree was bent over horizontal,' Banks says. Vegetation would have been flying around everywhere and smaller trees were likely to have been torn out and flung about in the gale. As others before Banks have noted, the Wollemi pine's exile to the canyon is a double-edged sword. 'The gorge where the trees are today,' he observes, 'acts as their protector and their destroyer.'

A team from Charles Sturt University has been studying the leaf structure of the pines and another group of scientists from the University of New South Wales has even isolated Wollemi pine essential oil, which evidently smells like Pine-O-Cleen.

Rod Peakall ran genetic tests on trees from the third stand and once again was unable to find any variation in the DNA.

The Wollemi Pine

To enter the ten-hectare Wollemi pine growing facility in Queensland you have to pass through a twenty-five-strand electrified fence, which packs a punch of 8000 volts, then step into two foot-baths, designed to kill any pathogens on your shoes. I arrived at Gympie on a steaming hot March day, without a breath of wind, which gave the place a tropical feel.

The centre is situated on a flat plain, at the end of a lonely road, and surrounded by pine forest. Many of its buildings are demountables. The only hint of the strange business under way was a glimpse of Wollemi pine foliage through an office window. Visitors are issued with daggy broad-brimmed hats, sunscreen and rubber boots, and many of the staff wore specially designed t-shirts, emblazoned with the Wollemi pine logo.

Malcolm Baxter, the manager of the Queensland Department of Primary Industries' nursery, led me into a large shaded growing area. 'This is where we bring people for the wow factor,' he said, with a proud sweep of his arm.

There, spread before me, was a carpet of green—Wollemi pines of every shape and size. At the entrance to that first growing area was a display showcasing the many different ways the tree could be grown. There were several types of bonsai, groundcovers, plants in hanging baskets and then the regular plants in pots, but even there the difference in the plants was mind-boggling.

In the wild there are fewer than 100 of these rare trees and yet in just one shade house tens of thousands were thriving. At the time of my visit in March 2005 there were several hundred thousand young trees growing at Gympie. The world has not seen such a forest of *Wollemia nobilis* for many millions of years.

Because of the strict security, the theme park atmosphere, and also because the species is now synonymous with dinosaurs, I couldn't help but think of *Jurassic Park*. There was, though, another reason for recalling the movie. In the Hollywood blockbuster, Dr Ian Malcolm responds to a puzzled question about how the island's dinosaurs could possibly breed considering they are all females.

'I am merely stating that uhh...life finds a way,' he replies.

Wollemi pines, I realised, were like Malcolm's dinosaurs but without the malevolence. *Wollemia nobilis* had found a way to escape from the very edge of the precipice that is extinction. How many critically endangered species have experienced a ten-thousand-fold increase in a decade?

The facility began its propagation with a small quantity of seedlings and cuttings, which Mount Annan Botanic Gardens had cultivated, ranging from tube stock to a few plants in thirty-centimetre pots.

The Wollemi pine growing facility was established as a partnership between DPI Forestry, a business group of

Queensland's Department of Primary Industries and Fisheries and Birkdale Nursery. Essentially the Queensland government grows the trees and Birkdale sells them. The profits are distributed and the New South Wales government earns a royalty on each sale, which is directed to help fund the conservation of the Wollemi pine and other threatened plant species.

As a teenager I had a pet axolotl. As soon as I left my bedroom and watched it through a tiny crack in the door it would swim around like crazy. No matter how quietly I tried to re-enter, however, the creature would freeze in mid-movement and drop to the bottom of its tank, motionless. I experienced a similar sensation encountering these Wollemi pines—it was as if they were *pretending* to stand still. They all looked so different: despite being the same age, all of the trees were a different height. Some had long straight branches, others were shaggy and droopy and some were multi-stemmed. How is it that a plant without any known genetic variability produces offspring as different from each other as individuals on a packed bus? Also at Gympie that day was Mount Annan horticulturist, Trish Meagher. 'This confirms the trees are not genetically identical,' Meagher said, one hand cupping a frond of a waist-high *Wollemia*. 'It's just that we haven't found the variability yet. They don't do anything you tell them to, they're very naughty. They're fast and loose and precocious.'

The Wollemi is also regarded as a more pleasant tree to propagate than hoop or bunya pines. 'It's much nicer to grow because it's not so prickly,' Baxter says.

Lyn Bradley is a senior member of Baxter's team and is one of the world's most passionate Wollemi pine fanatics. She is small but strong, with a cheeky smile that she constantly tries to tame, and she shares a great rapport with her staff. Bradley talks about the trees as if they were her high-achieving children, yet even among such a mass of greenery she has favourites.

'I'm sure everyone working here has their own plant marked out,' she says, 'and quite often we will hear somebody singing to their plant.

'The Wollemi pines continually surprise us,' Bradley continues. 'We call them our juvenile delinquents. They go through this nice reasonable-looking plant stage and then they hit puberty. They're like pimply teenagers and then they begin to blossom into these really elegant trees. But the only real similarity among them is that they're all green. They are a law unto themselves.'

There are around forty employees at the Wollemi pine facility, all of whom have undergone police checks and signed confidentiality agreements. Casuals are not allowed to work by themselves and contractors are always accompanied by a member of staff. During my visit Bradley tells me, 'We keep the project a secret, only the staff really know about it.' I was also surprised to discover that back in March I was the first journalist permitted to visit the nursery.

The process of demystifying Wollemi pines is also well under way, with staff developing their own non-scientific language for the tree—fronds are now referred to as 'feathers' and the resinous sap that the Wollemi pines use to protect their growth tips against frost are called 'polar caps'. These caps are likely to become one of the trees' signature idiosyncrasies, spectacular balls of gooey sap.

Lyn Bradley now has a sense of their annual cycle and has noted how their appearance changes dramatically with the seasons.

'In late May and June the older plants go into a dormancy period when they get their resinous, waxy polar caps,' she says. 'In late August–early September the waxy coating starts to soften and new shoots form. They get this stalk that only a mother could love—it looks like an asparagus stem. The new growth is a bronze colour and, amazingly, all of them lean towards the sun. I remember coming in one morning and thousands of stalks had all moved in the same direction.

'Once they start a growth spurt you can see up to fifty

centimetres of new foliage in a couple of weeks and a metre of growth easily in a year and people can't believe how soft and fern-like their leaves are.'

As the season warms, the bronze leaves turn to a soft green and then to the true blue-green colour of the rest of the tree—the adult leaves are as different from the juvenile as a tadpole is from a frog.

Bradley leads me to a quiet corner of the growing area where the truly strange Wollemi pines are kept. These freaks of *Wollemia* are know to those in the horticultural industry as the 'sports' and a couple are to be found in every 10,000. They are kept and grown on because no one knows when a strange mutation might turn up and become a nursery bestseller. So far none of the 'sports' pines appear to be anything more than scientific curiosities. One looks as if it has a pineapple growing on top, another is as prickly as a bunya pine and a third has flat leaves, without shoots, growing out of its trunk instead of from its branches.

One of the first tasks at Gympie was to build on the knowledge of the scientists in New South Wales and work out how to produce enough Wollemi pines to satisfy demand from around the world. At the heart of this effort are thousands of Wollemi pines, which have been designated as hedge stock.

These plants are fed lots of fertiliser so they continue to grow throughout the year, producing an endless supply of fresh tips, which can be cut off and planted.

'We milk the hedge plants all the time,' says Baxter. 'The strike rate used to be about 8 to 10 per cent but we have now been able to pull it up to about 95 per cent.'

Just how this is done is a well-kept secret. Thousands of cuttings are taken every day by the gangers and their staff. Teams amid the rows of Wollemi pine hedges continually carry armfuls of fronds to another team, which knows precisely which tips will turn into viable plants, and it then wraps these cuttings in wet hessian.

Bradley tells me they call the hessian 'nappies'. Within fifteen minutes of being harvested from the hedges the cuttings are further divided into segments. 'Within half an hour they are in a pot,' Bradley says. 'The sap that comes out during this process is so sticky you could just about use it as super glue.'

Chances are if you buy one of the first release Wollemi pines it will have been handled by 'setter' Carol Farkas. The cuttings are then taken to one of sixteen igloos, where they grow to the size of tube stock and are then put into pots. To keep up a steady supply of soil, a machine nicknamed the 'mutt' is used to fill the pots.

Lyn Bradley then walked me to the edge of a pit filled

with offcuts of the Wollemi pine that were never planted. This is where the 5 per cent that failed to strike roots and the even smaller percentage of trees that simply don't grow end up.

From the day news of the Wollemi pines made the papers, people have felt they would be a commercial hit simply because of their novelty value. What nobody could guess is that the trees would be so successfully cultivated. It seems *Wollemia* is about as tough as a plant can be without being a cactus.

Barbara McGeoch, owner of Wollemi Pine International, says the biggest concern was that the trees would turn out to be horticultural duds—impossible to propagate, delicate and too ecologically narrow to be popular across the horticultural spectrum.

'People want plants that are easy to grow,' McGeoch says. 'They want versatility and low maintenance. Although the Wollemi pine is endangered, it is also very tough.'

It seems extraordinary that one of the most critically threatened plants on Earth can be so robust. It has a high tolerance to hot, cold, sunshine, shade, low- and high-watering regimes as well as different types of soils. At this stage at least the Wollemi pine seems to be one of the most forgiving plants ever to enter the horticultural market.

'I can't believe their will to survive,' Bradley explains. 'If you look at the conditions they're actually growing in, in the wild in the canyon—low light, low pH, they're like a person who has been locked away for a few years.

'When it gets lots of love, light and water, it's like *Yee-ha!* They really are greedy little buggers.'

Wollemi project officer Kate Murray warns not to kill them with kindness. 'We water ours once a week. Don't keep them wet. They like to go a bit dry. *Never* sit them in water. Let them dry out but not to become powdery bone dry.'

The trees fare very well in pots, especially if left to grow on patios, and will survive for years in a smallish pot. Because of their striking form they are also likely to be popular with landscape designers as architectural plants in the garden.

The trees become dormant when kept inside. This ability to slow down their growth means they will hold their condition for many months, even away from direct light and fresh air. Indoor Wollemis do, however, need to be given spells outside and the horticulturists have found indoor plants do much better near doors and windows where they can get light and air. It is important to remember what these trees have to tolerate in the wild. A rainforest tree may sometimes resemble a young sapling but in fact be many decades old, biding its time for a break in the canopy before embarking on a growth spurt. For this reason even a human lifetime spent inside is probably of little challenge to a *Wollemia*.

'Don't be afraid to bring a Wollemi pine inside,' Murray says. 'But when you put it back outside, don't stick it into direct sun, give it time to adjust.'

They can be pruned heavily and, as the propagation method has shown, the trees survive even the toughest of hedging regimes. The absolute optimum conditions for the tree appear to be dappled sun, water once a week, use of a slow release fertiliser—ensuring it is sprinkled away from the stem, and pruning during winter only. They should be repotted every two years and indoor Wollemis should be given a week outside every month. And chances are that even if these guidelines aren't adhered to your tree will be just fine.

'You can give them anything at any time,' Murray says, 'and they just lap it up.'

On hearing what a super-species the Wollemi pine is proving to be in cultivation I couldn't help but wonder whether it has the potential to become a worldwide weed. This is a question John Benson from Sydney's Royal Botanic Gardens has also been considering. Only a fortnight before my visit to Gympie, Benson prepared a statement on the issue.

The Wollemi pine is available from quality nurseries and comes in a variety of sizes.

'Weed species,' he wrote, 'often have lots of small seeds that are easily dispersed. They have seeds with a long dormancy and a quick rate of growth after germination. Usually herbivores don't like to eat them. Finally they are super reproducers able to readily bud, grow from fragments or root suckers.

'Seed production in the Wollemi pine is low in its wild population. Most seeds disperse by wind but due to their weight it is unlikely they would disperse long distances,' he says. 'The seeds also have a limited dormancy and they are likely to become unviable after a few years.

'In ideal growth conditions the Wollemi pine grows about one metre per year and would take decades to grow into a large tree.'

'There's no history of other Araucariaceae becoming major weeds anywhere else in the world,' Benson later told me on the phone. 'However, you can't say whether in some places—a few places in the world—it is possible it could become a bit weedy. You can't rule it out, but given what we know it is unlikely and we won't know for fifty years anyway.'

At Gympie, the first batch of cloned Wollemi pines had already grown nearly knee-high at the time of my visit. Using this tissue culture technique means it is possible to produce a *Wollemia* from a few cells. Instead of having to rely on hedges and cuttings, mastering this method would mean being able to produce an endless supply of Wollemi pines. Unlike their cutting colleagues, cloned Wollemis all look the same—at least for the first few months. Though they too are beginning to break out out of their genetic straitjackets.

'Even the tissue cultured Wollemis are starting to assert their own personality,' Bradley told me as we stood beside a table groaning under the weight of hundreds of cloned Wollemi pines.

On my last afternoon at Gympie, Trish Meagher, Wollemi Pine International's marketing manager Sally McGeoch, and Kate Murray had an important job to do. In one of the shade houses was a collection of healthy and large Wollemi pines. Each of these trees was propagated from a known tree in the wild population—in the row where I was standing all were descendants of the Bill tree, otherwise known as King Billy. The provenance of the average *Wollemia* that will go on sale is not known—the job of keeping records for so many trees is too cumbersome. In that area, however, each plant had a known parent and could be sold as a premium product. The very first tree to be sold at auction by Sothebys in October 2005 will come from that collection.

An animated discussion was under way about what constituted a beautiful Wollemi pine: height, bushiness and an indefinable aesthetic. A code designating each tree's different qualities was allocated by the women: a score out of three for beauty; a mark for stature—waist-, breast- or greater than head-height; and a final code for being either single-stemmed, thin (having leading branches), or bushy.

The trio walked up and down the rows of Wollemi pines carrying clipboards, looking every bit like military commanders inspecting their troops. It was a surreal moment, a reality check for an endangered species about to be revealed to the world and which will now need to hold its own as a product. The job of assessing each of the several hundred plants was to take a couple of hours and soon I wandered off to a corner of this cultivated forest and crouched down until I was below the canopy.

I looked around and found a comfy-looking pot to sit on. It was a privileged moment. I had sat underneath the wild Wollemis deep in their canyon and had felt the power of ancient survival. Now I was with a group of *Wollemia* youngsters, soon to make their debut. It was a little like sitting among a crowd of extras waiting to rush on to a stage or a movie set. As I peered closer at the green foliage I saw signs of life everywhere—ants, a lizard and a lady beetle. I remembered Bradley telling me her staff had found tree snakes, carpet snakes, red-bellied black snakes and frogs all trying to live in this stand of Wollemi pines. Everywhere all around me life was trying 'to find a way'.

Look hard at a Wollemi pine and before long you start thinking about the wonder of existence—how a tree unknown to science a little over a decade ago, and on the brink of extinction, could make such a comeback.

Acknowledgments

This book has taken some time to come together and, in the process, a number of other stories have developed from it. In early February 1999 I did a story on the bunya pines. Prue Bartlett, the ranger responsible for the trees, and I fell in love in her bunya pine forest while we talked about Wollemis. Prue has encouraged me, read my drafts, suggested changes and listened to hours upon hours of me raving on about a tree. My beautiful boys Angus and Finn have been patiently curious about their Dad writing a book about the 'dinosaur trees'.

Fiona Inglis, my agent, was a source of advice and help through a lengthy process. I thank Michael Heyward for editing the book; his ability to sort through the meanderings of the early versions has won my utmost respect.

This book would not have been possible without the research of palaeobotanists such as Bob Hill, Mike Macphail, Peter Kershaw, Wayne Harris, Liz Truswell, Jane Francis, Imogen Poole, Greg Jordan and Alan Partridge. I am also grateful to the teams at the New South Wales National Parks and Wildlife Service and Sydney's Royal Botanic Gardens who have spent the last five years studying the Wollemi pine; John Benson and Ken Hill both took countless calls and assisted me with much information; Cathy Offord, Graeme Errington and Patricia Meagher welcomed me many times to their lab and greenhouses at Mount Annan; Carrick Chambers supplied me with a great deal of information about Koonwarra; Brett Summerell shared his knowledge of the tree's microscopic companions and foes; David Crust and Michael Sharp told me about their experiences at the end of a cable and Crust guided me in the wilderness. I have drawn heavily on the work of all of these people and all gave me access above and beyond what I expected. I also thank David Noble, Haydn Washington and Jan Allen for sharing their stories with me and Wyn Jones for showing me *Microstrobos*. Rod Peakall was enormously patient as he helped me understand plant genetics.

Others who helped include Tom Rich, Alex Ritchie, Jim Allen, Mike Archer, Barbara Briggs, Walter Boles, Bob Conroy, Andrew Cox, Anna Dawson, Allan Hansell, Geoff Hope, Frank Howarth, Mark McGruther, David Marr, Lisa Morrisett, Max Moulds, Keith Muir, Peter Prineas, Lin Sutherland, Andy Thomas and Mary White.

Bibliography

NEWSPAPER AND MAGAZINE ARTICLES

Archer, Mike, 'A Legacy of Boats, Bloats and Floaters', Australian Museum, *Nature Australia*, vol. 26, no. 3, pp. 70–71.

Benson, Simon, 'Wollemi Duo Barred from "Their" Pine', *Daily Telegraph*, 4 July 1997.

Jones, Cheryl, 'Triumph of the Coneheads', *Bulletin*, March 3, 1998, pp. 34–35.

Macdonald-Leigh, Sharon, 'Career Path', *Sydney Morning Herald*, 18 July 1998.

Macphail, Mike, et al, 'Wollemi Pine—Old Pollen Records for a Newly Discovered Genus of Gymnosperm', in *Geology Today*, March–April 1995, pp. 48–50.

Vickers-Rich, Patricia and Rich, Tom, 'Australia's Polar Dinosaurs', in *Scientific American*, July 1993, pp. 40–45.

Woodford, James, 'Found: Tree from the Dinosaur Age, and It's Alive' and 'A Chance Discovery Unveils Hidden Gorge's Age-Old Secret', *Sydney Morning Herald*, 14 December 1994.

'Science's Reluctant New Botanic Star', *SMH*, 17 December 1994.

'Alive in Tasmania: A Giant Tree 8000 Years Older than Christ', *SMH*, 28 January 1995.

'Fears for Ancient Pines after Scavengers Find Secret Grove', *SMH*, 24 February 1995.

'Tree Chic and Cloned: The Ancient Wollemi Pine', *SMH*, 4 May 1995.

'Hopes Rise As More Rare Pines Found', *SMH*, 6 May 1995.

'Dinosaur Plants Ready to Sprout Million-Dollar Sales', *SMH*, 12 December 1995.

'Secret Canyon Reveals its Buried Treasure', *SMH*, 20 April 1996.

'Botanical Find of the Century Yields Anti-Cancer Drug', *SMH*, 18 June 1996.

'The Jurassic Tree and the Lost Valley', *SMH*, 7 June 1997.

'Giant Bunya Bombshells', *SMH*, 12 February 1999.

Wroe, Steve, 'Killer Kangaroos and Other Murderous Marsupials', in *Scientific American*, vol. 280, no. 5, May 1999.

JOURNAL ARTICLES AND ABSTRACTS

Andrews, Alan, 'Draftsman D'Arcy's Colo River Surveys 1831–1835', in *Journal of the Royal Australian Historical Society*, December 1979, pp. 173–187.

'Govett's Luck: Assistant Surveyor Govett and the Southern Tributaries of the Colo River', in *Journal of the Royal Australian Historical Society*, March 1978, pp. 263–269.

Benson, John and Redpath, Phil, 'The Nature of Pre-European Native Vegetation in South-Eastern Australia: A Critique of D. G. Ryan, J. R. Ryan and B. J. Starr (1995) *The Australian Landscape—Observations of Explorers and Early Settlers'*, in *Cunninghamia* 5(2) December 1997, pp. 285–328.

Chambers, Carrick, Drinnan, Andrew and McLoughlin, Stephen, 'Some Morphological Features of Wollemi Pine (*Wollemia nobilis*: Araucariaceae) and Their Comparison to Cretaceous Plant Fossils', in *International Journal of Plant Science*, 159(1), 1998, pp. 160–71.

Gilmour, Simon and Hill, Ken, 'Relationships of the Wollemi Pine (*Wollemia nobilis*) and a Molecular Phylogeny of the Araucariaceae', in *Telopea*, December 1997.

Gross, Janet, Rich, Tom and Vickers-Rich, Patricia, 'Dinosaur Bone Infection', in *National Geographic Research and Exploration*, 9(3), 1993, pp. 286–93.

Harris, Wayne, 'Basal Tertiary Microfloras from the Princetown Area, Victoria, Australia', in *Palaeontographica*, April 1965, pp. 4–6, 75–106.

Hill, Ken, 'Architecture of the Wollemi Pine', in *Australian Journal of Botany*, vol. 45, 1997, pp. 817–26.

'The Wollemi Pine: Discovering a Living Fossil', in *Nature and Resources*, vol. 32, no. 1, 1996, pp. 20–25.

Hill, Robert and Brodribb, Tim, 'Southern Conifers in Time and Space', in *Australian Journal of Botany*, vol. 47, 1999, pp. 639–96.

Hill, Robert and Scriven, Leonie, 'The Angiosperm-Dominated Woody Vegetation of Antarctica: A Review', in *Review of Palaeobotany and Palynology*, vol. 8., 1995, pp. 175–98.

Hogbin, Patricia, Peakall, R. and Sydes, M. A., 'Achieving Practical Outcomes from Genetic Studies of Rare Australian Plants', in *Australian Journal of Botany*, vol. 48, 2000.

Jones, Wyn, Hill, Ken and Allen, Jan, '*Wollemia nobilis*, A New Living Australian Genus and Species in the Araucariaceae', in *Telopea*, September 1995, pp 173–76.

Kershaw, Peter, Moss, P. and Van der Kaars, S., 'Environmental Change and the Human Occupation of Australia', in *Anthropologie*, vol. xxxv, no. 2–3, 1997, pp. 35–43.

Macphail, Mike, 'Impact of the K/T Event on the South-East Australian Flora and Vegetation: Mass Extinction, Niche Disruption or Nil?', in *Palaeoaustral*, vol. 1, 1994, pp. 9–13.

Offord, Cathy, Meagher, Patricia, Porter, Carolyn and Errington, Graeme, 'Sexual Reproduction and Early Plant Growth of the Wollemi Pine (*Wollemia nobilis*), A Rare and Threatened Australian Conifer', in *Annals of Botany*, vol. 84, 1999, pp. 1–9.

Peakall, Rod, 'Exceptionally Low Genetic Diversity in an Ancient Relic, the Wollemi Pine: Implications for Conservation Theory and Practice', Abstracts, Genetics Society of Australia, 45th Annual Conference, Sydney 1998.

Pickett, J. W., Macphail, M., Partridge, A. and Pole, M., 'Middle Miocene Palaeotopography at Little Bay, Near Maroubra, New South Wales', in *Australian Journal of Earth Sciences*, vol. 44, 1997, pp. 509–10.

Strobel, Gary, Hess, W., Li, Jia-Yao, Ford, Eugene, Sears, Jo, Rajinder, Sidhu and Summerell, Brett, '*Pestalotiopsis guepinii*, a Taxol-Producing Endophyte of the Wollemi Pine, *Wollemia nobilis*', in *Australian Journal of Botany*, vol. 45, 1997, pp. 1073–82.

Truswell, E. M. and Marchant, N. G., 'Early Tertiary Pollen of Probable Droseracean Affinity from Central Australia', in *Special Papers in Palaeontology*, no. 35, 1986, pp. 163–78.

Wroe, Steve, 'Estimating the Weight of the Pleistocene Marsupial Lion, *Thylacoleo carnifex*', in *Australian Journal of Zoology*, vol. 47, 1999, pp. 489–98.

Books, Reports and Other

Archer, Mike et al, *From Plesiosaurs to People: 100 Million Years of Australian Environmental History*, from State of the Environment Series, Department of Environment, Canberra, 1998.

Riversleigh: The Story of Animals in Ancient Rainforests of Inland Australia, Reed, Sydney, 1991.

Benson, John, 'Threatened by Discovery: Research and Management of the Wollemi Pine *Wollemia nobilis* Jones, Hill and Allen', in *Back from the Brink: Refining the Threatened Species Recovery Process*, S. Stephens and S. Maxwell (eds), Surrey Beatty & Sons, Chippendale, 1996.

Boland, B., Chippendale, H., Johnston, H. and Turner, K., *Forest Trees of Australia*, CSIRO, East Melbourne, 1992.

Bowman, David, *Australian Rainforests*, Cambridge University Press, Cambridge, 2000.

Colong Foundation for Wilderness, National Parks Association, *The Wollemi Wilderness Plan*, December 1997.

Darwin, Charles, *The Origin of Species*, first pub. 1859, 6th edn, J. M. Dent & Sons Ltd, London, 1971.

Enright, Neal and Hill, Robert, *Ecology of the Southern Conifers*, Melbourne University Press, Carlton, 1995.

Flora of Australia, Ferns, Gymnosperms and Allied Groups, volume 48, AGPS, Canberra, 1998.

Fortey, Richard, *Life: An Unauthorised Biography*, HarperCollins, London, 1997.

Hill, R. S., Truswell, E., McLoughlin, S. and Dettmann, M. E., 'The Evolution of the Australian Flora: Fossil Evidence' in *Flora of*

Australia vol. 1, 2nd edn., A. E. Orchard (ed.), ABRS/CSIRO, Canberra, 1999, pp. 251–320.

Hill, Robert, *History of the Australian Vegetation: Cretaceous to the Present*, Cambridge University Press, Cambridge, 1994.

Kershaw, Peter, et al, 'Patterns and Causes of Vegetation Change in the Australian Wet Tropics Region over the Last 10 Million Years', in *Patterns and Causes of Vegetation Change in Tropical Rainforests*, Chicago University Press, Chicago, 1999.

Kirkpatrick, Jamie and Backhouse, Sue, *Native Trees of Tasmania*, Pandani Press, Hobart, 1981.

Seventeenth International Congress of Genetics, *Genetics and the Understanding of Life*, Birmingham, 1993.

Silverman, Milton, *Search for the Dawn Redwoods*, monograph, self-published, date unknown.

Studdy-Clift, Pat, 'Brigands of the Bush', *Stockmans Hall of Fame*, Longreach, March 1993, pp. 8–9.

The Lady Bushranger: The Life of Jessie Hickman, Hesperian Press, Perth, 1996.

Washington, Haydn, 'The Colo: A Wilderness Won?', in *Fighting for Wilderness*, G. Mosley and J. Messer (eds), Fontana/ACF, 1984, pp. 23–45.

Weinberg, Samantha, *A Fish Caught in Time: The Search for the Coelacanth*, Fourth Estate, London, 1999.

White, Mary, *After the Greening: The Browning of Australia*, Kangaroo Press, Sydney, 1994.

The Greening of Gondwana, Reed, Sydney, 1986.

'A Green Dinosaur', videotape from science series 'Quantum', ABC TV, Sydney, 1997.

Index

Other Bestsellers by James Woodford

THE SECRET LIFE OF WOMBATS

WINNER, WHITLEY AWARD FOR BEST POPULAR
ZOOLOGY BOOK

In 1960, a fifteen-year-old schoolboy called Peter Nicholson
began to investigate the secret world of wombats by crawling
down their burrows and making friends with them. These
torchlight adventures significantly expanded the field of
wombat studies. In *The Secret Life of Wombats*, James Woodford
pursues Nicholson's story and embarks on his own adventures
to uncover the true nature of our most intriguing marsupial.

'A surprisingly great read…Outstanding.' Tim Fischer

'Woodford has done the research, he has read widely, spoken
with the major wombat pundits and with the lay observers.
He has travelled to gain direct experience of all species.
His book has drawn together much of what is known about
wombats…I know more about wombats than I did, and retain
some stark images which I hope never to lose.' *Sunday Age*

'This book will change our view of one of the world's most
intriguing and intelligent marsupials.' *Wild*

Non-fiction, paperback, 240pp, illus, rrp$24.00,
ISBN 1 877008 43 5

THE DOG FENCE
A JOURNEY ACROSS THE HEART OF AUSTRALIA

At 5400 kilometres, the Dog Fence is one of the longest man-made structures on Earth. It slices across Australia's desert heart, dividing the continent to keep dingoes away from livestock. This is James Woodford's enthralling account of his journey over sand, gibber plains and salt lakes. *The Dog Fence* is about the hazards of travel, the lessons of history and the passion and resilience of Australian men and women on the land.

'For anyone who loves the Australian bush, this book is irresistible.' *Courier-Mail*

'A thrilling account of a little known Australia, this is a funny, warm and compelling tale of the outback, bushmen and wildlife.' *Age*

'Woodford's cast of characters and his raw material vary with the bewildering richness of the Australian landscape and the fence line on the author's city eyes is sufficiently intense to make the book very much his own.' *Australian*

'A graphic and compelling account of one of the world's most bizarre yet least-known structures, the dingo fence.' *Sydney Morning Herald*

Non-fiction, paperback, 272pp, illus, rrp$24.00,
ISBN 1 920885 26 9